PURSUING
SUSTAINABILITY

PURSUING
SUSTAINABILITY

A Guide to the Science and Practice

PAMELA MATSON,

WILLIAM C. CLARK, AND

KRISTER ANDERSSON

PRINCETON UNIVERSITY PRESS

Princeton and Oxford

Contents

Preface

This book provides a general framework to help people interested in mobilizing knowledge to promote sustainability. We wrote it to complement the many excellent books, courses, and programs that are focused on specific aspects of sustainability: works on the use of land and water, the production of energy and food, corporate responsibility, the challenges of governance and conflict resolution, the pursuit of poverty alleviation and equitable growth. Our goal is not to supplant those more focused efforts but, rather, to sketch a broad perspective on sustainability that places those efforts in a broader context, with an appreciation for the interrelationships among them. The book could therefore be used as a companion or background reading for courses on topics such as sustainable energy, agriculture, or urbanization. It could also serve as a short guide or primer for students and practitioners at all levels who are seeking a more systematic and comprehensive platform on which to base their own pursuits of sustainability.

To these ends, we have tried to write this book as if it is part of a conversation with our readers over important issues in sustainable development, rather than as an argument that we want to win. This has meant keeping the text as short and readable as possible, leaving out many important topics. It has meant, due to our collaboration, addressing a broader sweep of the vast range of material relevant to sustainability than any of us could have done alone; but within that range, we have focused on telling a decidedly interdisciplinary story. Our goal of preparing an accessible primer or companion guide has led us to avoid extensive citations in favor of suggestions for a limited number of "further readings" at the end of the book. It has also involved a constant battle against disciplinary jargon—a battle that we know we haven't altogether won. An extended glossary of essential terms, linked to places where those terms are first introduced in the text, is our attempt to limit the resulting damage. We hope that the result of our efforts, combined with your own personal passions, knowledge, and experience, will help you join us in nudging the world along a transition toward sustainability. All our children, grandchildren, and generations beyond are counting on it.

Acknowledgments

The ideas in this book reflect our own personal learning through decades of individual research, teaching, and outreach efforts, but most especially they reflect what we have learned from a number of close colleagues. Among the great many people to whom we give our thanks we owe a special debt to Partha Dasgupta, Bob Kates, Bill Turner, Lin Ostrom, John Schellnhuber, and John Bongaarts. With them, we have tried in large and small ways to formulate a clear statement of what is needed for a transition to sustainability. To them, we say thank you for the inspiration, ideas, and encouragement to keep at it.

Many others have motivated and inspired us. We especially thank Peter Vitousek, Anni Clark, Carla Andersson, Stephen Carpenter, Roz Naylor, Ruth DeFries, Ganesh Shivakoti, Danny Lam, Diana Liverman, Julia Novy-Hildesley, Banny Banerjee, Theo Gibbs, and our children, Liana and Michael Vitousek, Graham and Adam Clark, and Markus Andersson.

The book has benefited from the advice of many reviewers. To them we give all our thanks (and assign no blame). Kai Lee, Ruth DeFries, and Kimberly Nicholas provided incredibly useful in-depth reviews of the first draft, and Kai and Kim went on to provide equally high-quality comments on the second. Noelle Boucquey, Andy Lyons, Jesse Reeves, Alan Zarychta, Kelsey Cody, and many students in our undergraduate classes at Stanford, Harvard, and University of Colorado wrestled with and commented on the book in its early drafts.

The various versions of this book could not have been completed without the intellectual and logistical contributions of Noelle Boucquey and Peter Jewett. To them we give our deep thanks for helping make the book a reality.

Over the years, many funders supported the research and actions that led us to this book, and we thank them all. For the book writing, we are particularly thankful for partial support from Italy's Ministry for Environment, Land and Sea, and, of course, our universities.

PURSUING
SUSTAINABILITY

CHAPTER 1

Pursuing Sustainability: An Introduction

Sustainability is a term we hear all around us. Corporations brand themselves as sustainable and attempt to build sustainability goals, measures, and metrics into their business plans and supply chains. State and local governments set sustainability targets and pursue them, installing efficiency standards and practices, encouraging the use of public transportation systems, and incentivizing citizens to reduce, reuse, and recycle. Universities compete for sustainability awards that recognize efforts ranging from improving energy and water use efficiency to curricular offerings. Researchers focus their attention on the development of new knowledge and technologies to promote sustainability. Consumers consider sustainability concerns as they buy organic, or buy certified sustainable seafood or wood products. Citizens strive to reduce their environmental footprints on the planet out of a sense of responsibility to their children and grandchildren.

Sustainable development, likewise, is a widely used term. It frequently appears in high-level discussions of the United Nations, the World Bank, and non-governmental organizations (NGOs) such as CARE and WWF and is a fundamental objective of the European Union and of many nations rich and poor. The World Business Council for Sustainable Development counts many leading global companies among its members. These and similar organizations invest in efforts to help countries, companies, and communities "develop" not just in the near term but for the long-term benefit of people.

While the terms *sustainability* and *sustainable development* are often used by different communities of people, the vast majority of these uses have something very important in common: a realization that our ability to prosper now and in the future requires increased attention not just to economic and social progress but also to conserving Earth's **life support systems**: the fundamental environmental processes and natural resources on which our hopes for prosperity depend. Because of that commonality, we use the terms interchangeably. We believe that the take-home messages of this book are important for both.

THE EVOLUTION OF SUSTAINABILITY THINKING

Sustainability is an old idea. Societies for centuries have recognized the importance of demanding no more of the environment than it can supply over the long term. This recognition is evident in long-standing ideas about fallowing fields and conserving game and protecting water sources. The concept of sustainable development in its modern form, however, was most famously articulated by the United Nations World Commission on Environment and Development (WCED) in 1987. As Gro Brundtland, the commission chairwoman, wrote at the time: "Environment is where we live; and development is what we all do in attempting to improve our lot within that abode. The two are inseparable."[1] The commission argued that *sustainable* development "meets the needs of the present without compromising the ability of future generations to meet their own needs." And its carefully documented report left little room for doubt (among those willing to confront the evidence) that a transition toward sustainable development would have to arrest and reverse the increasingly global and accelerating degradation of Earth's environment and natural resources.

Since the Brundtland Commission's call for global action the UN has reinforced the urgency of a **sustainability transition** through numerous conferences and agreements. Over many years and several iterations, the international community has developed consensus on a wide-ranging set of sustainable-development goals,[2] and in pursuit of those goals, numerous efforts at the global, national, and local levels have been launched. They aim to reduce hunger and poverty; improve access to health care, family planning, and education; increase agricultural production while reducing environmental degradation; and halt the degradation of Earth's life support systems. Today, thousands of governmental and non-governmental organizations, private firms, and individuals all over the world have adopted the idea of sustainability and have started to allocate attention and resources to sustainable development programs. Corporations have developed metrics to track the economic, social, and environmental impacts of their actions. Cities around the world have created and joined associations designed to share best practices and encourage progress. Regional and national efforts are to be found in every part of the world. Scholarly organizations such as the world's national science and engineering academies and numerous professional scientific associations have also engaged in the effort. And international consortia such as the World Business Council for Sustainable Development and the Science and Technology Alliance for Global Sustainability have built cross-sectoral communities of actors from private and public organizations to chart strategies for progress toward sustainability goals. (We provide in appendix B a short list of the Internet resources we have found most useful for keeping abreast of rapidly changing initiatives around the world that are pursuing sustainability.)

As a result of these many initiatives, people's vision of sustainable development has been rapidly evolving. It has matured from simple relationships that see human prosperity primarily in terms of economic growth, to a focus on human needs as called for by the Brundtland Commission, to the ever more encompassing and nuanced views centered on social well-being being advanced today (see chapter 2).

A benefit of this expanding framing of sustainability has been the creation of an ever-broader "tent" under which the multiple constituencies of scholarship, advocacy, and action are now working to promote sustainable development. This inclusiveness, however, creates an increasing risk of losing sight of the sustainability forest for its many individual trees and the concomitant temptation for everyone engaged to fall back into activities focused on individual disciplines or sectors. Progress toward sustainability requires commitment to such details—to the individual trees of our global system as well as the particular contexts in which they must be nurtured if they are to contribute to sustainable development. But progress also requires a broader perspective on how the parts of the forest depend on one another, interact, and co-evolve. We have tried to provide one such perspective with this book.

SCIENCE FOR SUSTAINABILITY

Meeting the challenges of sustainable development requires action that changes the status quo. Because too many powerful interests have a stake in resisting such change and in continuing "business as usual," meeting the challenges of sustainable development will require a deep and broad commitment to social agitation, to stirring things up. Contributions to this agenda of agitation and action are needed not just from self-described activists but also from political and business leaders, civil society, medical professionals, educators, and individual citizens. In addition, however, sustainable development requires contributions from *scientists*, a term we use to encompass all sorts of scholars committed to figuring out how the world works—natural and social scientists, humanists, policy analysts, engineers, medical scientists, and all their many kindred.

Just what are the roles for science and scientists in promoting sustainable development? We explore many specific contributions throughout this book. In general, however, the roles of science include helping society to see where present trends are taking us, to discover or design new technologies and policies that might change our course, and to evaluate the possible trade-offs and implications for future generations of implementing such alternatives. Or, in the words of Nobel Laureate Amartya Sen, the role of science is to help assure that the social agitation seeking to promote sustainable development is *informed* agitation.

When we, the authors of this book, first began working on sustainability issues, lots of relevant science and technology were being done, but there were few places in which those who were doing them could escape their disciplinary homes to come together in collaborations centered on sustainable development. Today, things have changed. Like the fields of health or agricultural science before it, **sustainability sci-**

ence has emerged as a field focused on creating and harnessing many different kinds of knowledge to help address social problems. In particular, it is a field that tackles directly the numerous problems involved in pursuing sustainable development. To that end, sustainability science strives to integrate study and practice through use-inspired research. It includes the contributions of many different kinds of basic science and the contributions of people involved in the design and implementation of policy, technology, and practices. The field also carries out the unique and critical role of striving to integrate those knowledge bases and to build on them in new ways, developing knowledge to support decision making for sustainability goals. Ultimately, sustainability science is about increasing our knowledge of and ability to manage the interactions between environmental and social systems that set the stage on which sustainable development plays out.

We have a good deal to say later in this book about the general characteristics of such **social-environmental systems** and of the common challenges people face as they guide those systems along paths toward sustainable development. To bring specificity to those generalized discussions, however, we have found case studies useful. We introduce in the next section of this chapter four such studies that represent the kinds of challenges facing sustainable development and the types of work sustainability scientists do in helping to meet those challenges.

SUSTAINABILITY CHALLENGES IN THE REAL WORLD: FOUR CASE STUDIES

As noted in the previous section, we have selected four case studies to illustrate just how challenging it is to determine what to do to promote sustainability over the long term. The cases take place in a wide variety of settings—high-tech science laboratories, fields of small-scale farmers, neighborhoods of one of the world's most cosmopolitan cities—but they share some central features. The cases are alike in that they showcase people trying to make good things happen for themselves and the people in their communities, struggling to deal with unintended consequences, and continuing to push for progress despite failures and setbacks. The cases show that even the most well-intentioned interventions can go awry for many different reasons, including not considering them in the context of the full social-environmental system, not building appropriate and useful knowledge and getting it into the hands of the right decision makers, not working effectively within governance systems that allow sustainable decisions, and not having good luck. Each case has unique lessons to share, and together they give us concrete experiences through which to think about sustainability concepts.

These cases form a starting point for this book. We draw from them in all chapters of the book as we share our perspectives on what we think are the most critical lessons for those of us who are pursuing sustainability. We present short introductions to each case next and provide more detailed treatments in appendix A. We believe that most readers will find it helpful to read the fuller treatments in the appendix before getting too far along in the book.

London

The London case highlights the challenges of thinking analytically about development over the multigenerational time periods central to concerns about sustainability.

London today is widely recognized as a leading world city. It regularly scores at or near the top of surveys about sustainability and quality of life in urban areas, racking up especially high scores for its international clout, its technological savvy, and its livability; doing well on economy, governance, and many dimensions of environment; but still struggling with air pollution and social inclusion.

This has not always been the situation for London. On the contrary, London's current high but uneven prosperity rests on a history of more than a thousand years of collisions with the environment, some very like those being experienced in today's rapidly growing "new" cities. Two related themes have run through these collisions, both relevant to cities everywhere.

One involves the constant struggle to secure the basic food, fuel, and material resources needed for a growing city and at the same time to dispose of the waste products resulting from the use of those resources. Failures to handle these resource and waste flows directly resulted in multiple episodes of food shortages and consequent malnutrition for the poor, the intermingling of human waste and drinking waters, chronic air pollution, and growing vulnerability to flooding.

The second recurrent theme in the relationship of London to its environment centers on its population's battle with communicable disease. The odds in this battle were clearly affected by repeated failures to solve the resource management issues noted previously, but they also involved the dynamics of resistance and immunity in a closely packed settlement increasingly exposed to diseases imported from remote corners of the world.

In grappling with these problems, growing numbers of Londoners pursued short-term personal gains that repeatedly inflicted long-term large-scale costs of social and environmental degradation that eventually became untenable. But with each setback, society responded with a mix of political activism, scientific discoveries, technological inventions, social adjustments, and new forms of governance that—together with events occurring in the wider world beyond London—opened the way for the next round of development initiatives. These invariably led to their own surprises and readjustments, the most recent or persistent of which are being publicly debated in the context of the London Plan, an organic document that charts the city's goals and strategies for a sustainability transition.

Nepal

The Nepal case focuses on the multiple pathways through which irrigation technology has been deployed to reduce food insecurity. It highlights the importance of involving technology users in the design and operation of technology-augmented systems.

Growing food in Nepal is a challenging prospect. With most of the country mountainous, and area for agriculture very limited, Nepali people have always been susceptible to food shortages. By the late twentieth century, the growing rural population was becoming chronically food insecure. One possible solution that had been successfully pursued elsewhere was to introduce improved irrigation technology to increase the amount of food produced on land already under agriculture. But when the Nepali government and foreign partners began to construct advanced irrigation systems to replace the more primitive version being used by poor farmers, the results fell short of expectations. Many of the agricultural systems that received high levels of financial support and technological improvements actually saw declining food production.

What was going on here? How could these state-of-the-art engineering projects not deliver improvements to farmers in Nepal? Scientists studying the irrigation projects found the answer in key elements of farmers' social systems that the "outsider" aid providers had overlooked. Many of the systems were designed in government engineering offices and failed to incorporate local farmers' knowledge about typical stream flows and other aspects of local context. In particular, because the sophisticated equipment used to create the new irrigation works was difficult to maintain and operate, the government agencies took over from farmers full responsibility for the new irrigation systems. Farmers themselves were thus no longer in charge of maintaining these systems, as they had been for the earlier simpler ones. As a result, farmers perceived little need to cooperate with other farmers to coordinate their irrigation activities. Upstream farmers stopped worrying about maintaining good relationships with downstream farmers, withdrawing more and leaving less water for their downstream neighbors' use. Eventually, agricultural productivity within the system as a whole declined.

Over the years, researchers studying irrigation activities in Nepal have shown the important role that irrigators themselves play in making irrigation systems work. They fulfill several important functions, such as organizing contributions of labor to the construction and maintenance of field canals and regulating water allocation. And they perform the critical functions of monitoring violations and helping enforce compliance measures. Their long-term commitment to their village as well as their knowledge of the place, its resources, and its people makes an important difference in the performance of irrigation systems. However, even farmer-managed systems can benefit from expertise, experience, and financing from outside organizations. Current efforts to introduce new technologies to improve food security in Nepal are thus increasingly experimenting with hybrid models involving significant roles for both farmers and outsiders.

Yaqui Valley

The Yaqui Valley case takes a closer look at how imperfect but persistent interdisciplinary research on the coupled social-environmental system of a region can bring tangible benefits for sustainability when conducted in close consultation with local decision makers.

The Yaqui Valley, located in the state of Sonora in northern Mexico, was the home of the Green Revolution for wheat. It is where, starting in the 1950s, the international agricultural research community developed the new high-yielding cultivars of wheat and corn that have been critical to meeting the food needs of people around the world. Farming in the valley has benefited from that new knowledge and know-how, and farmers there boast some of the highest yields of wheat in the world.

Unfortunately, the valley today has a range of sustainability challenges, most of them unintended consequences of the Green Revolution. For example, water resources are used very inefficiently (through flood irrigation in the fields), and until recently, the irrigation districts lacked rules for changing water draws to sustain water resources during times of drought. The agricultural systems tend to be overfertilized, and the addition of far more nitrogen than is used by crops leads to nutrient losses (in the form of greenhouse gases, and air and water pollutants) from soils to neighboring streams, ocean, and atmosphere.

If a "breadbasket" is not developing sustainably, what is to be done? Over a fifteen-year period, an interdisciplinary research team sought ways to reduce the environmental impacts of agriculture in the valley while maintaining or even improving grain production and economic well-being. Doing so required a focus on agronomic and environmental as well as economic, social, and political aspects so as to understand the coupled social-environmental system of the Yaqui Valley and to come up with solutions that made sense in that context.

In the case of overfertilization, researchers and farmers worked together to develop and test alternative fertilizer practices that required less fertilizer input, lost less fertilizer to water and air systems, and kept grain amounts and quality high. These and other win-win management practices were shown to be capable of saving both farmers' money and the environment. But discovering great win-win technologies didn't mean they would be used. Ultimately, it also took an understanding of the broader decision-making and governance systems of the valley, and improved partnerships with farmers and credit unions, to turn good ideas into action on the ground.

Stratospheric Ozone

The ozone case highlights the challenges of managing useful technologies that turn out to pose threats to the global commons. It shows how careful orchestration of international scientific assessments and stakeholder engagement, coupled with a willingness to make decisions in the face of uncertainty, produced a successful global environmental regulation.

The mechanical refrigerator, invented in the mid-1800s, allowed foods and other materials to be kept cold under all conditions. Not surprisingly, this invention was a boon to human health and well-being. The main problem with it, however, was that its standard cooling fluid was ammonia, a gas with a dangerous propensity to explode! Thanks to the work of industrial chemists, a new class of industrial chemicals—called chlorofluorocarbons, or CFCs—was invented in the 1930s to replace the dangerous refrigerant. After testing, CFCs were judged to be safe, nontoxic, and nonexplosive. Their manufacture and use took off, and CFCs became an important ingredient in a variety of coolants, and later, also propellants and cleansers. By all accounts, the class of compounds was an amazingly useful technology that could help meet some of the needs and wants of people around the world.

In the 1970s scientists discovered—much to their surprise—that CFCs were accumulating in the upper atmosphere, or stratosphere, and could be found all across the planet. Moreover, by the 1980s, researchers who would later win the Nobel Prize in Chemistry had shown that CFCs have the capacity to deplete ozone in the stratosphere. This finding was important because the *ozone layer* is known to serve as a shield that protects life on Earth from health damages that would otherwise result from excessive exposure to ultraviolet (UV) radiation from the sun. People and other forms of life around the planet were at risk.

What happened to prevent humans, and all the other species who share the planet, from being "fried" by UV radiation as the ozone layer was lost? Thanks to ongoing measurements and monitoring and scientific analysis, and the leadership and engagement of decision makers in governments, corporations, and civil society, the ozone problem was identified and addressed through a sequential set of international agreements. These agreements limited and then radically reduced the use of ozone-depleting substances, and ultimately replaced those substances with others that have fewer negative consequences. While the loss of stratospheric ozone has not been completely reversed, the decline has been arrested, and signs of a recovery are evident.

THE BIG THEMES OF THIS BOOK

Our short case studies suggest (and their longer versions in appendix A show conclusively) that pursuing sustainability is not easy. That making progress is difficult is confirmed by the unabated emissions of greenhouse gases and their consequences for people and ecosystems, by the appalling conditions in many of the world's largest and most rapidly growing cities, and by the deepening inequalities in prospects for improved well-being of individual humans around the world.

Progress is nonetheless happening in some places and sectors, and is possible elsewhere. We now know that the successful pursuit of sustainability almost always requires the participation of many different stakeholders, including those of us in the "knowledge production world" of universities, research institutes, technology innovation operations, and policy think tanks. No one person can be an expert in all the fields of knowledge that may prove essential in solving particular sustainability problems. Thus, the successful pursuit of sustainability is usually context specific and a collaborative affair. But there are a few things that our own experience suggests should be understood by *everyone* who wants to be an effective contributor to the pursuit of sustainability. That essential foundation of general knowledge about sustainable development is what we have tried to convey in the remainder of this book.

In chapter 2 we present a framework for thinking about sustainability—a framework that links sustainability goals to their ultimate determinants. In particular, we follow a growing body of scholarship and practice in arguing that the ultimate goal of sustainable development should be focused on human well-being. Moreover, we propose that well-being should be thought of inclusively; assuring the well-being of a few people today should *not* be achieved by degrading the well-being of their neighbors or their grandchildren. We suggest that the ultimate determinants of inclusive well-being should be thought of as the stocks of assets on which people now draw and will draw on in the future to subsist and to improve their lives—stocks that include natural, social, human, manufactured, and knowledge capital. Chapter 2 also explores the status of sustainable development today in terms of measures of inclusive human well-being and of the asset base that supports well-being.

Chapter 3 suggests that pursuing sustainability is complicated by the fact that efforts to do so play out in rather challenging ways in the social-environmental systems that humans inhabit. These are not simple equilibrium systems. Instead, they are complex, dynamic, adaptive systems that involve a great variety of interactions among social and environmental components, with all kinds of trade-offs and surprises in store. Focusing on a single component—for example, a new technology, a certain tropical rainforest, a particular type of material or pollutant, or particular new policy—in isolation from the rest of the system will rarely result in successful problem solving. Diagnosing problems and developing, implementing, and evaluating solutions requires an understanding of the system as a whole.

Developing knowledge of how coupled social-environmental systems work is necessary for making progress toward sustainability, but it is not sufficient. It is also necessary to understand how people, as active, committed agents of change, can intervene in those systems to make them work differently. Almost always, such interventions require collaboration to be effective. The governance processes that are the focus of chapter 4 are how society secures such cooperation in the face of conflicting goals, temptations to free ride, and the prevalence of outright selfishness. Governance systems, working at multiple levels, influence when and how society takes actions that promote or undermine sustainable development. Chapter 4 aims to characterize the nature and importance of governance processes and to give some practical guidance to those trying to reform such processes in ways that support the effective pursuit of sustainability.

New knowledge, tools, and approaches are also crucial determinants of sustainability. Unfortunately, however, even really innovative and exciting ideas cannot promote sustainable development if they are not actually used by decision makers. Chapter 5 discusses the challenges of creating useful knowledge and of linking such knowledge with action. It outlines approaches for increasing the chances that innovations will be useful and used effectively in decision making for advancing sustainable development.

Finally, in chapter 6, we try to tie the big ideas of previous chapters to their implications for individual people as agents of change. What should these big ideas mean to each of us? How can each of us, as individual agents, contribute to the pursuit of sustainability? What are the attributes individuals need to serve as leaders in a transition to sustainability? What type of training can help prepare such leaders? What can institutions of learning do to help? What should one remember when working in the world to promote sustainability?

* * *

The conceptualization of sustainability, the framing of sustainability goals, and the perspectives on meeting the challenges of sustainable development that we present in this book are certainly not the only approaches worth considering. (We list some different approaches we have found especially thought provoking in the "Additional Resources" section of appendix B.) Rather, they simply constitute the most useful framework that we three coauthors have been able to construct from our diverse experiences as professionals who both study and engage in efforts to promote a sustainability transition. We encourage our readers to push back on us when their values or experiences lead them to conclusions different from ours. We nonetheless hope that many of you will find in this short book some building blocks that can contribute to your own pursuit of sustainability.

CHAPTER 2

A Framework for Sustainability Analysis: Linking Ultimate Goals with Their Underlying Determinants

Given the complexity of the sustainability challenge, how can progress be made? Is there a way to conceptualize the multiple dimensions of sustainability so that important elements and outcomes are recognized and not ignored? Can the challenge of sustainable development be framed in such a way that allows us to keep eyes open to complications and interactions and still move ahead toward sustainability goals? In this chapter we sketch a framework for thinking about sustainability we have found to be helpful.

Building on the work of our colleague Partha Dasgupta and a growing number of scholars and international organizations,[1] we summarize here our preferred framework for analyzing sustainable development, elaborate on it in the remainder of this chapter, and develop it throughout the book.

> Development is sustainable if inclusive social well-being does not decline over multiple generations. Well-being is achieved through the consumption of goods and services that are produced as part of the dynamics of social-environmental systems. Well-being is ultimately grounded on five clusters of underlying capital assets: stocks of natural, manufactured, human, social, and knowledge capital. If the aggregate capacity of these assets to generate value for society is allowed to degrade over time, social well-being will ultimately de-

cline as well, and development will be unsustainable. Policy to promote sustainability is ultimately about how to manage these assets, and people's access to them, so that neither the social wealth they represent nor the social well-being they create declines with time.

Let's expand on this rather dense summary using the organizing structure of figure 2.1. At the top of the figure are the goals of sustainability and sustainable development. For reasons explained in the next section, our framework follows current practice in expanding the Brundtland Commission's sustainability goal of "meeting people's basic needs" with a more expansive goal focused on their *well-being*. What people view to be the most important **constituents of well-being** will vary depending on individual circumstances and value systems. For most people, however, the important constituents include capacity to meet their basic needs for food, water, shelter, energy, and physical security. Many would add to this list factors that make life not just livable but good: access to health, education, nature, community, and opportunity. But sustainable development has always been about not just individuals but all humanity. Thus, our interest focuses on the aggregate property that we call **inclusive social well-being**: *social* because it is meant to be more than just the sum of individuals' well-being, and *inclusive* because of the concern for equity within and between generations. Although our goal for sustainable development is characterized in terms of inclusive social well-being, we'll abbreviate it as *well-being* in much of this book. We discuss well-being in greater depth in the next section of this chapter.

At the bottom of figure 2.1 is the base or foundation of **capital assets** that constitute the ultimate determinants of well-being. Taken together, these assets are the "fuel in the tank" that powers development and on which society draws to create its well-being. The framework we present here characterizes a society's assets in terms of five component "capitals": natural, manufactured, human, social, and knowledge. A short description of each of these terms is presented in table 2.1. We expand on those descriptions later in this chapter, where we also discuss the links among the capital assets and their collective significance for well-being and sustainability. In particular, we argue that under certain circumstances inclusive well-being will track with an aggregate

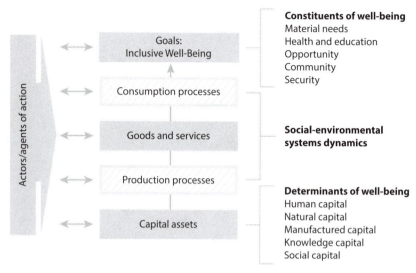

FIGURE 2.1. A framework for understanding and pursuing sustainability. See the text for a full description.

measure of these capital assets. That measure, which has come to be called "inclusive (social) wealth," is now being used regularly to monitor progress toward sustainability.

In the middle of figure 2.1, connecting well-being with the capital assets that are its ultimate determinants are the complex dynamics of the social-environmental systems. We focus on those dynamics in chapter 3. For our purposes here, however, it is useful to think of social-environmental systems in terms of their production and consumption processes. In this context, **production processes** can be seen as transforming capital assets into goods and services relevant to sustainable development. Production of food, electricity, housing, information technologies, and the accompanying wastes are examples. **Consumption processes** use goods and services created by production processes to achieve (or not) the goals of sustainability (i.e., in our case, inclusive social well-being).

Some would argue that sustainability challenges are largely a problem of consumption. We believe, however, that production and consumption should be viewed as parts of an integrated social-environmental system. Consumption demand does indeed drive production, but production supply can shape consumption by creating demand where none

TABLE 2.1. CAPITAL ASSETS
Key Elements of Capital Assets and Some of the Scholarly Fields That Study Them

Capital Asset	Key Elements or Characteristics	Associated Fields of Study
Natural capital	Land, water, biotic, mineral resources; climate and atmosphere; biodiversity; etc.	Geography, Earth systems science, ecology, conservation biology, natural resources, ecological economics
Human capital	Human population (size, distribution, health, education, other capabilities)	Demography, health and medicine, education, labor, geography
Manufactured capital	Buildings (homes, factories and their products); infrastructure (transport, energy, information)	Industrial ecology, green design, pollution control, sustainability engineering, geography
Social capital	Laws, norms, rules, customs; institutions (political, judicial, economic); trust	Political economy, institutions, policy, government, sociology, law, geography
Knowledge capital	Codified knowledge (conceptual, factual, practical, know-how)	Policy studies, innovation and design, science and technology studies, social learning, geography

existed, for example, for personal computers and, later, smartphones. From a sustainability perspective, both consumption and production processes can be part of the problem or part of the solution.

Today, consumption varies widely depending on the region or socioeconomic class to which a person belongs. Some people in some areas of the world need to consume more—they need access to food, water, energy, clothing, and shelter, as well as education and health care, if their well-being is to improve. For example, the consumption of additional protein sources by undernourished children and adults can make a huge difference in health and mental development. In contrast, overconsumption has become a health and lifestyle challenge in some regions, with negative impacts on social well-being as well as the natural environment. Total global consumption is on the increase. To meet this demand, production processes are placing demands on key assets—such as fuel, soil, minerals, and water—in greater quantities than ever before.[2]

As production processes strive to meet rising consumption demands (and to encourage new ones) they likewise can be conservative or profligate in their use of resources and their effects on the environment. *How* goods and services are produced, as well as *what kind* and *quantity* are consumed, can have a huge influence on whether, and how quickly, society can navigate a transition toward sustainability. To date, much of the progress toward sustainability goals has been associated with new technologies and policies that reduce the negative impacts of production processes—think of the development of clean-energy technologies, of no-till agriculture practices, or of the treaties resulting in the phase out of dangerous chemicals that are described in our case study of the stratospheric ozone layer.

Increasingly, however, attention is also being paid to reforming consumption. Initiatives ranging from promoting energy and water conservation to encouraging less consumption of red meat are gaining headway around the world. A central challenge for those pursuing sustainability is finding ways to ensure that both the consumption of goods and services necessary to promote human well-being and the production processes necessary to supply them are pursued in ways that generate well-being while their negative consequences for social-environmental systems are reduced.

An example of that challenge is provided by our case study on agricultural production in the Yaqui Valley. The capacity to meet the consumption needs for food and nutrition of a rapidly growing world population in the second half of the last century was ultimately dependent on new production processes related to photosynthesis and farming. The Green Revolution was all about meeting growing needs for food consumption by developing new technologies that produced more cereals—wheat, maize, and rice, the staple foods for much of the world's population. The new production technologies succeeded in meeting consumption needs, but they inadvertently led, in many places around the world, to loss of biodiversity, overuse of water, and overuse of fertilizer, with attendant air and water pollution and greenhouse gas emissions, and even displacement of social groups, all with consequences for long-term social well-being. But, as our case study suggests, many of those negative consequences resulted from the way the production processes were carried out, and many have been or could now be modified to reduce the negative outcomes. In the coming decades, the world has to avoid as many such harmful consequences of production as possible and, in some cases, to reduce the need for production by reducing excess or inefficient consumption and waste.

The Yaqui Valley example illustrates nicely an added complication with which efforts to pursue sustainability must deal: the production-consumption processes that connect capital assets to human well-being cannot be understood as merely economic systems of supply and demand, any more than they can be understood as merely systems of resource depletion and regeneration. Rather, as suggested by the middle panel in figure 2.1, production and consumption processes are embedded in the dynamics of the complex, interacting social-environmental systems that constitute the stage on which initiatives to promote sustainable development must be carried out. We postpone until chapter 3 a deeper consideration of those dynamics—and the challenges and opportunities they provide in the pursuit of sustainability.

A final feature of our framework for analyzing sustainability is captured in the left-hand-side panel of figure 2.1, which reminds us of the importance of **actors** and **agency**. The system linking sustainability goals through production and consumption processes to underlying capital assets is not a mechanical clock or computer model that, once started, just plods on its inevitable path to the future. Rather, it is a system in which choices made by various actors and agents of change matter. The "choosers" may be individual people or research teams or advocacy groups or provider firms or governments. They may be selfish or public spirited, or well informed or ignorant. None of them are likely to be sufficiently powerful or brilliant to make the whole social-environmental system conform entirely to their individual wishes. Most, however, will find that their choice of actions matters for the pursuit of sustainability and that their choices matter more if they can carry out those actions in cooperation with others. We discuss the pathways open to actors and agents of change in the pursuit of sustainability in chapters 4 ("Governance") and 5 ("Linking Knowledge with Action").

* * *

While the framework captured in figure 2.1 is conceptual rather than quantitative, the perspective is increasingly being used to evaluate the sustainability of alternative development pathways.[3] We find the framework useful because it reminds us to consider and evaluate multiple dimensions of the sustainability challenge and, when interventions are planned, to look for more than simple cause-effect relationships. The rest of this book is devoted to unpacking the components and relationships of figure 2.1 and to discussing how the resulting understanding can be used to pursue sustainable development. Our first step on that journey is a closer look at the meaning of well-being.

CONCEPTUALIZING WELL-BEING

What is well-being? How should we think about it and its implications for action? These and similar questions of value have deeply engaged a variety of ethical and religious traditions throughout history. Answers are invariably and properly personal, reflecting individual histories and contexts. The approach we have found most useful for pursuing sustainability has several broad features that transcend individual circumstances.

First, we adopt an anthropocentric perspective. Our definition of sustainable development in terms of well-being is ultimately about *people*. Environment matters too: in fact, radically improved stewardship of the environment is a necessary condition for sustainable development. But sustaining the environment is not inherently the same as sustainable development. In the framework presented here, environmental conservation matters as a means for advancing human well-being and thus sustainable development, rather than as an end in itself. That said, there is room in our approach for people who want to put a high value on nature not only as an instrumental source of the food they eat and water they drink but also as a foundation of the spiritual dimensions of their well-being, or because they value nature in its own right.

Second, we have a broad view of human well-being. Our conceptualization builds on broader philosophical discussions about "the quality of life" that far predate the idea of sustainable development.[4] The constituents of well-being, which we will discuss in more detail, can include people's access to basic material goods and security but also to health, education, fulfilling social relations (community), enjoyment of nature, or the opportunity to make choices about the direction of one's life.

Third, as noted earlier, we focus on "inclusive social well-being," which consists of the aggregated well-being of people across relevant intervals of space and time. Which intervals? Spatially, we could be talking about the well-being of people included in a household, community, region, country, or the world; to be specific, we need to define that society and always be aware of its connections to others and ultimately to the global system. Temporally, we aggregate well-being across multiple generations, not just one.

Fourth, we recognize that relating *individual* well-being to *inclusive social* well-being raises profound ethical questions. To evaluate particular development pathways with respect to their sustainability involves grappling with questions such as how to decide and who gets to decide what constitutes individual well-being, as well as examining the implications for equity of how a society chooses to define and measure inclusive social well-being. (Box 2.1 provides a few examples of the "environmental justice" problems that result when concepts of well-being are narrowly focused on individual communities.)

Finally, we acknowledge that by defining sustainable development in terms of inclusive social well-being that is merely "nondeclining" across generations we may seem too modest in our ambitions. What of development pathways that might lead to an ever-improving record of well-being? What of "optimal" development? In response, we can say only that though we would applaud such developments, our ambitions are indeed more modest. (But so were those of the Brundtland Commission in its original focus on development that "meets the needs of the present without compromising the ability of future generations to meet their own needs.") Why? Because we find compelling enough the ethics of leaving our grandchildren at least as much as we inherited from our grandparents, and hard enough the challenge of figuring out how to do it.

In our perspective on sustainability, short-term needs do not automatically trump the needs of people in the longer term; intragenera-

BOX 2.1. Environmental Justice and Inclusive Social Well-Being

What happens when the needs of one person or community interfere with the ability of other people in other places to meet their own needs? What happens when those other people already have the lowest level of well-being? This is the crucial issue addressed in efforts toward "environmental justice." Born from recognition of the fact that low-income and minority communities across the globe are more likely to live in polluted conditions and near hazardous industries (e.g., mines, textile factories, waste dumps, refineries), the environmental justice movement fights for legislation and practices that protect communities from harm.[5]

Examples of injustice abound. At the global scale, for example, environmental justice problems occur when industrialized countries ship their waste to less developed areas of the world. For example, people in China working in and living near recycling and disposal facilities for electronic waste, which is often shipped from the United States, are exposed to hundreds of toxic chemicals.

Within-country problems are even more well known but are also easier to address, especially in industrialized countries. Let's use an example from the United States. In the early 1980s, a hazardous waste dump was constructed in Warren County, North Carolina, about an hour northeast of the Raleigh-Durham area. The county was largely rural, poor, and African American—notably different from the wealthy suburbs of the state capital. When residents learned of the wastes to be deposited in their soils—from which PCBs and other chemicals would potentially leach into the groundwater—they staged six weeks of protests by marching and lying in the streets to block dump trucks. Though the people's lack of power with the state legislature meant they ultimately lost the fight, their story galvanized a wider environmental justice movement in the United States.

Unfortunately, similar stories originate in many other parts of the world, and today, environmental justice struggles continue. Nonetheless, important progress has been made. In the United States, for example, federal agencies have been required since 1996 to identify and address any disproportionate impacts of their policies on low-income and minority communities. In the South African Constitution, finalized in 1996, an included bill of rights grants South Africans the right to an "environment that is not harmful to their health or well-being" and the right to "ecologically sustainable development."[6] Likewise, the European Union has issued declarations that all people have a right to a healthy environment, most recently in 2000 with the "Charter of Fundamental Rights of the European Union."[7]

tional well-being does not necessarily trump intergenerational well-being. Likewise, the needs of one person or community do not trump the ability of other people in other places to meet their own needs. Our own personal values put particular weight on improving the well-being of those who have least of it (rather than merely striving for average well-being) and on securing the well-being of people across generations extending out at least to the grandchildren of today's youth. But other people will have other values. We therefore tried not to hardwire our own values into this book. Rather, we have worked to present a general structure that invites readers to make explicit which values regarding well-being and equity they want to incorporate into their analyses and advocacy of sustainability and sustainable development.

CONSTITUENTS OF WELL-BEING

How do people experience well-being? Well-being is certainly multidimensional, but there are common elements that figure strongly in many assessments.

In everyday language, people might use the term well-being to refer to a certain level of health, happiness, or success in life. But what level of health should be expected? How is happiness defined? What activities or achievements constitute success? Are there universal aspects of well-being, or do they differ across regions and cultures? What do humans need for survival, and what are reasonable aspirations? One way to discover answers to these questions is simply to ask people through national and global opinion surveys.[8] These provide subjective information about people's levels of well-being and happiness and can give information about the relationship between people's subjective well-being and variables such as income or level of basic needs fulfilled.

Researchers from diverse fields have also focused on this question. Most agree that the core of well-being rests in a combination of material, social, and personal fulfillment. One definition created for practitioners of development policy recognizes well-being as something that is achieved "when human needs are met, [and] when one can act meaningfully to pursue one's goals."[9] Others have included, along with basic needs, personal security, strong social ties, health, autonomy, and vocational satisfaction.[10] Nobel Prize–winning economist Amartya Sen suggested that the conceptualization of well-being should go beyond fulfilling "needs" to celebrating and expanding human "capabilities" as "agents of change who can—given the opportunity—think, assess, evaluate, resolve, inspire, agitate, and, through these means, reshape the world."[11]

For this overview, we focus on six constituents of well-being that appear frequently in everyday speech and have also emerged as the result of research. In the following paragraphs we define these constituents, briefly discuss global trends for each, and examine a few of the ways in which they interact with the underlying capital assets that are their determinants.

Material Needs

People have different notions of needs (versus wants or luxuries), but at the most basic level, humans need food, water, energy, and shelter to survive. When these material needs are satisfied, they provide a foundation of well-being from which to engage in personal or professional growth.

In much of the world, basic material needs are being met more comprehensively than they were even a few decades ago. For example, food consumption per person per day rose from a global average of about 2400 kilocalories (kcal) in 1970 to almost 2900 kcal in 2010, made possible in part by advances in agricultural production like those discussed in our Yaqui Valley case study. The distribution of calories consumed, however, is not equal across regions or socioeconomic groups. In fact, while people in industrialized countries are consuming up to 3400 kcal per day, the average per capita consumption in sub-Saharan African nations is about 2200 kcal per day (with some people getting much less).[12] Despite the abundance of food in the world today, an estimated one in eight people still suffer from chronic hunger.[13] In sub-Saharan Africa, for example, hunger persists owing to weak governance institutions, global market failures, and persistent gender and socioeconomic inequalities that reduce people's access to food; the natural capital to supply these needs is probably sufficient in many places but not being effectively accessed.

Meeting the material needs and wants of people requires the use of capital assets. Natural capital (like water resources, crops and agricultural lands, fossil fuels, minerals, air, and climate) is drawn on (or otherwise affected) to produce the goods and services that people need and want. The processes of producing and consuming items to meet needs and wants of people can have more or fewer negative effects on natural capital, depending on how they are carried out. For sustainable development, progress will have to be made toward supporting the material needs of people without placing such a large and unsustainable burden on natural capital. Improved knowledge capital (for example in the form of improved crop varieties and better understanding of nutrition) and social capital (for example in the form of international agreements, property rights, economic policies, and cultural mores) will almost certainly be needed to move development along more sustainable paths.

Health

Health is probably the most universal and immediately recognizable component of well-being. Without good health, an individual's quality of life suffers, making it more difficult for individuals and communities to reach their potential or other dimensions of well-being. Strong human capital comprises healthy people, and they are thus a determinant as well as a constituent of well-being. Several indicators suggest that across the globe, dramatic improvements in human health have long been under way.

One of the simplest and most dramatic indicators of the improvements is life expectancy at birth. Other things being equal, if people are living longer, they have the potential for experiencing more well-being. For most of human history, life expectancies at birth hovered around 30 years, though much less healthy places and periods (such as the London of the mid-eighteenth century described in our case study) can readily be found. Beginning in the middle of the eighteenth century, however, life expectancies began to rise dramatically: people born in today's healthiest countries can expect to live almost three times as long as most of their ancestors. This trend began in northern Europe but now is observed in all parts of the world, though the amount of progress remains very uneven among different regions (figure 2.2).

Other dimensions of health have improved as well. In eighteenth-century London, more than a third of children born died as infants.

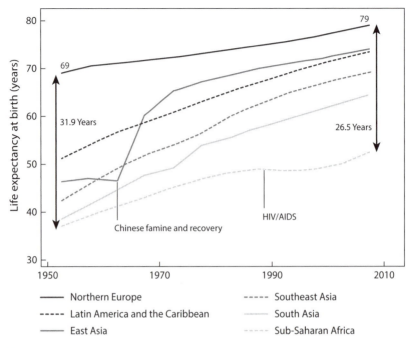

FIGURE 2.2. Life expectancies at birth, in years, by region, 1950–2010. (Deaton, Angus, *The Great Escape*. © 2014 Princeton University Press. Reprinted by permission of Princeton University Press)

Today, with better nutrition, vaccines, and medical care, the number of deaths of children under the age of five has declined from 12.4 million in 1990 to 8.1 million in 2009. The number of women dying from pregnancy and childbirth complications also decreased by nearly 50% between 1990 and 2013.[14] Ironically, with the aging of populations in developed countries, a major shift in global disease is under way, away from the infectious diseases common to developing countries and young populations toward the chronic diseases (or lifestyle diseases) of developed countries and aging populations (box 2.2).

As with the satisfaction of material needs, health outcomes are highly disparate around the globe, with some regions bearing very intense health burdens. For example, in 2015, a child born in sub-Saharan Africa was twelve times more likely to die before age five than a child born in high-income countries.[15] Maternal health also varies widely

BOX 2.2. Global Trends in Disease and Mortality

Not long ago, the majority of deaths around the world resulted from communicable diseases, nutritional deficiencies, and childbirth complications. These problems were focused in the developing countries of the world, and global health and development communities put great effort into their solution. Ironically, just as these diseases are declining, global rates of noncommunicable diseases are on the rise.[16]

As of 2010, deaths from noncommunicable diseases accounted for two out of every three deaths globally, with heart disease and strokes alone accounting for one in four deaths (compared with one in five in 1990).[17] Sometimes termed "lifestyle diseases" because of their association with unhealthful diet and lack of physical activity, these diseases are already prevalent in developed countries, so developing regions are now experiencing most of the increases in suffering from these diseases.

Increases in cardiovascular disease and diabetes are particularly emblematic of this trend. Though heart disease contributes to a larger share of deaths in developed countries than in developing countries (49% versus 23%), in absolute numbers there are several million more people with heart disease in developing regions.[18] Moreover, while cardiovascular-related deaths have been falling since the late 1960s in developed regions, such deaths are increasing in developing areas, and disproportionally affecting younger people there. For example, in India, over half of cardiovascular deaths occur before age seventy, compared with less than a quarter in developed regions.[19] Likewise, diabetes prevalence is increasing in every region and is closely tied to rising rates of obesity. Globally, the number of people living with diabetes is expected to more than double from year 2000 numbers to 366 million in 2030,[20] with rates of increase projected to be substantially higher in developing regions.

The growing prevalence of lifestyle diseases is due in part to increasing urbanization as well as to longer average life spans. As more people move to urban areas and adopt sedentary habits, they become more susceptible to lifestyle diseases. For example, a higher rate of cardiovascular disease in urban Delhi compared with that in the surrounding countryside was also associated with higher levels of blood pressure, cholesterol, and diabetes.[21] Higher urban rates of diabetes have also been found in a variety of other regions, including Asia, Africa, and the Middle East.

Regardless of a person's urban or rural location, inadequate exercise and changes in diet that increase the proportions of satu-

---

BOX 2.2. (*continued*)

rated fat, sugars, and processed foods are associated with diseases like diabetes and cardiovascular disease. Tobacco use, also
associated with urbanization and industrialization, is increasing in
many regions and is a key factor in cardiovascular disease and
deaths.

In thinking about these shifting global disease trends, it is important to keep in mind regional variations. Though cardiovascular
disease and diabetes are increasing globally, HIV/AIDS, respiratory
infections, and diarrheal diseases are still the leading causes of
death in the lowest-income countries.[22] Further, a growing problem
in middle-income countries is a "double burden" of both high rates
of childhood communicable disease or undernourishment combined with rising adult rates of lifestyle diseases; approximately
40% of the world population is estimated to be living in areas subject to both types of risk factors.[23] And within developed nations,
socioeconomic disparities contribute to very different patterns of
health between those on the highest and lowest rungs of the economic ladder.

---

between developing and developed regions. Most pregnancy- and
childbirth-related deaths occur in developing countries and could be
largely avoided with better medical care.[24] In many areas facing poverty, multiple issues (e.g., with water, sanitation, access to medical infrastructure) combine to make it difficult for people to maintain good
health for themselves and their children.

In the confluence of factors that affect people's health, it is clear that
all five capital assets are involved. Human capital, of course, includes
healthy people, thus giving health the interesting character of being
both a constituent and determinant of well-being. Natural and manufactured capital can positively affect human health (e.g., clean water,
food, shelter, medicine, energy for cooking and heating), and the character of social and knowledge capital in different regions often affects
this relationship. For example, relatively simple hand-washing techniques and supplies—created and tested using manufactured and
knowledge capital—can be brought to even the poorest communities
with adequate social/institutional capital to facilitate the transfer.
Where health can be fostered, a positive feedback loop can also result:
the healthier a population is, the more effective its human capital can

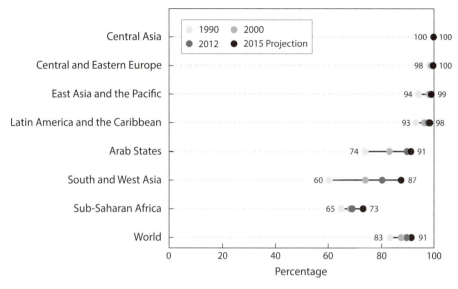

FIGURE 2.3. Change in youth literacy rates since 1990 and the projections for 2015. *Note*: Regions are sorted by the projected literacy rate in 2015. Data for 1990 refer to the period 1985–1994; data for 2000 refer to the period 1995–2004; data for 2012 refer to the period 2005–2012, and data for 2015 are projected. (UNESCO Institute for Statistics, www.uis.unesco.org/datacentre)

be at contributing to economic development, creating new knowledge, as well as contributing to social-environmental problem solving . . . and human health.

Education

As a component of well-being, education is the foundation for individual self-improvement as well collective advancement. Education is what allows people to take advantage of their experiences and the reservoir of knowledge capital in the world.

With education, the story is similar to that of other measures of well-being: gains are being made globally, but some regions and groups are being left behind.[25] Youth literacy has improved in all regions of the world, with the most significant gains seen in South and West Asia and the Arab states (figure 2.3). The global average for adult literacy has risen more than 20% since 1970, to 84% in 2012. However, this statistic masks the fact that one in five women worldwide remain illiterate, and millions of children are still not in school. The main obstacles to increasing education opportunities include violent conflicts, fragile governments, costs (to families and states), and ill health (of students and family members).

How is education linked to capital assets that are the ultimate determinants of well-being? The recognition that learning is a "key function in the creation of human capital"[26] makes education a priority for sustainable development. Social capital plays a critical role in that it includes the institutional structures that allow educational programs to be developed, funded, and delivered. The role of social capital in maintaining peace also permits education to take place. Knowledge capital, likewise, plays a crucial role as people create innovations in and for education around the world.

Opportunity

Opportunity refers to the ability of people to make choices about how they want to live and what they want to do. Opportunity is a key constituent of well-being because of its implications for pursuing a range of choices in life. A lack of opportunity may be experienced in a variety of ways. For example, a minority student in Chicago or Baltimore may experience a lack of opportunity because of underfunded schools, unsafe neighborhoods, and the continuing legacy of racism in America. A woman in rural India may experience a lack of opportunity because she has to spend hours each day procuring fuel and preparing food for her family.

Inequality is a major barrier to opportunity and is pervasive in much of the world. Currently, the richest 1% of the world's population have about half of all global wealth; the richest 10% have almost 90% of the wealth.[27] Likewise, within countries, the gap between the wealthiest and the poorest is widening. The groups most likely to suffer from wealth inequalities are indigenous peoples, minorities, and women.[28] Especially concerning is the persistence of these inequalities over generations, highlighting the lack of opportunities for people to escape the circumstances into which they are born.

Despite many challenges, people are working to improve opportunities in a variety of creative ways. Microfinancing programs aimed at women—a form of social capital—have helped provide them financial support to start small businesses that produce and use manufactured capital. In Bangladesh, policies aimed at empowering women led to declines in fertility rates and increases in life expectancies and childhood survival rates.[29] Natural capital (for example, in the form of a stable climate or availability of water or land for growing crops) and manufactured capital that allow people to access and use social and biophysical resources are also critical to the pursuit of opportunities.

Community

Like opportunity, the concept of community can be difficult to define. Particularly in today's interconnected world, the geographic, social, and political boundaries of a community can be porous and shifting. Nevertheless, in social-environmental systems, communities are generally understood to include all social system interactions within a defined geographic space[30] and have some level of "shared fate" (e.g., in experiencing a hurricane or earthquake). Community is an important constituent of well-being both in terms of people's experiencing the everyday strength (or weakness) of social ties and trust as well as the more long-term ability to be resilient in the face of social or environmental disruptions. Communities that cultivate strong human and social capital (including trusted governance institutions and good communication between different community groups) are better positioned to address such crises.

Security

Security is another important constituent in our analysis of well-being. The United Nations Development Programme (UNDP) defines human security as the "freedom from want and fear."[31] In this sense, human security encompasses more than just peace between nations; it includes individual freedom from crime, discrimination, and hunger.

Crime, civil conflicts, and terrorism are some of the largest threats to human security in today's world. Daily insecurity is experienced in regions with high crime rates. Within these regions, those groups already on the margins of society because of poverty or ethnicity often suffer the most. Reducing crime through increasing equality and additional means will have positive effects on the other constituents of well-being, allowing people to take advantage of knowledge capital and al-lowing human capital to develop further. Similarly, diminishing violent conflicts around the globe improves the ability of governments to man-age their natural capital resources, and citizens to take advantage of social capital.

DETERMINANTS OF WELL-BEING:
THE CAPITAL ASSET BASE

In our framework for sustainable development, inclusive social well-being is supported by the use or consumption of a broad range of goods and services, all of which are produced, drawn from, and ultimately determined by the capital assets of the planet. These capital assets can thus be usefully thought of as the "state variables" of the social-environmental system. The way in which they determine, and in turn are shaped by, production and consumption processes is the crux of sustainable development. In the following sections, we discuss in greater detail five groups of capital assets, taking stock of their chang-ing conditions, exploring their interactions with one another, and sketching the ways in which they are connected to inclusive social well-being.

Natural Capital

Natural capital is the stock of resources and environmental conditions provided by the Earth system, used to meet the basic needs of all peo-ple. It includes the atmosphere and climate, mineral resources, ecosys-tems and biodiversity, biogeochemical cycling, soils for growing crops, the crop plants themselves, groundwater or surface water resources, materials to build with, ocean fisheries, and so many other sources of the goods and services needed by humans. Humans could not live on Earth without access to these things, and indeed, *Homo sapiens* did not rise as a species until the stocks of many of these (for example, atmo-spheric oxygen levels!) were exactly right and could provide the global life support system humanity now enjoys.

Some kinds of natural capital have been recognized and valued by humans because of the goods and services derived from them, including, for example, the fuels and fibers and foodstuffs that people are accustomed to paying for. But much of the social value of natural capital has been unnoticed or undercounted by humans throughout history, because the goods and services produced from it are taken for granted. Why? Because they are "free": societies have never paid for them, and they are recognized only when they are lost or damaged (for instance, as discussed in our case studies, the protection provided by the natural capital of the stratospheric ozone layer). Today, researchers are working to develop a comprehensive accounting of the goods and services produced from natural capital to better understand the social value (monetary and otherwise) of conserving that capital. The less well-understood and accounted-for goods and services derived from natural capital include pollination of crop plants by wild and managed insects, prevention of erosion and recovery of eroded soils by certain kinds of plants and ecosystems, protection for fish nurseries provided by coral reefs and mangrove systems, provision of tourism and recreation opportunities as well as cultural and ethical values, and many more. And then, of course, there is natural capital that most humans would rather not have—mosquitoes are one good example!

What is the condition today of Earth's natural capital and the goods and services it provides? A series of assessments, beginning with the Millennium Ecosystem Assessment[32] and continuing with the work of the Intergovernmental Platform on Biodiversity and Ecosystem Services,[33] suggest that many different ecosystem services are in decline (table 2.2), yet these are goods and services future generations will surely need. Human activities are altering many different elements of natural capital (see figure 2.4 for a selection), and this decline in or degradation of the asset base undermines inclusive well-being and presents barriers to sustainable development.

Let's take a look at freshwater resources as an example of a natural capital stock. Direct consumption of clean freshwater is essential to human health, but freshwater is also a critical ingredient in manufacturing processes, energy production processes, and agricultural production and food processing, among others. Unfortunately, stocks of freshwater (a natural capital asset) are often depleted in the production of these goods and services. Groundwater is in many places being drawn

Earth's climate system is another critical and often overlooked component of natural capital. Earth's climate has changed over geologic timescales owing to, among other factors, natural changes in Earth's orbit around the sun, differences in solar intensity, and biological and physical changes in Earth processes. Nonetheless, over the past several million years, Earth's atmosphere and climate have remained in ranges tolerable for the emergence of human life, and over the past 10,000 years (following the last glaciation), climate has become remarkably consistent, supporting the establishment of agriculture and the rise of human societies.

In the last couple of hundred years, however, the scale of human activities has grown to the point that they are now changing the planet's climate. Consumption of fossil fuel energy (an energy asset that results from decomposed plant and animal material produced millions of years ago) and, to a lesser extent, use of land assets for livestock, agriculture, and forestry are releasing greenhouse gases into the atmosphere (see the top row of figure 2.4). Because of their radiation-absorbing characteristics, these gases trap heat in the lower atmosphere and thus warm the planet, leading to an increase in average global temperature (with greatest warming occurring in the high latitudes). As a consequence, ice sheets and glaciers are shrinking, sea ice and high-latitude spring snow cover are declining, sea level is rising, and ocean water is becoming more acidic. Many other effects are associated with climate change—from heat waves and increased intensity of storms to impacts on agricultural production and freshwater resources. All these changes put at risk the well-being of future generations.

Overall, natural capital plays an essential role in providing the underlying conditions for human well-being. The recent UNDP *Inclusive Wealth Report* noted that "a large part of what nature offers is a necessity and not a luxury. . . . caution must be taken [to prevent] irreversible processes that might cause a decrease in well-being."[34] Because many of the goods and services derived from natural capital are taken for granted and treated as though they were free, society tends to deplete or degrade them too readily and to invest too little in their maintenance and restocking.

Natural capital also interacts with other capital assets. It underpins much of manufactured capital (e.g., through the use of fossil fuels and construction materials to create manufactured capital) and supports healthy human capital (e.g., by providing clean air and water). The state of social capital in turn affects natural capital (e.g., natural resources often suffer in regions with weak regulations). Moreover, natural capital is affected by socioeconomic and political arrangements. Though many poorer countries are rich in natural resources, for example, the benefits of these resources often flow to wealthier nations in the form of raw materials or ecological services. Natural capital is thus deeply intertwined with human health, human activities, and human institutions, and its ability to provide for future well-being depends on those.

Manufactured Capital

A second asset stock of the social-environmental system is **manufactured capital**. This asset class includes human-made factories and the products they produce, transportation systems, dwellings, farming and water purification technologies, and energy infrastructure as well as the objects—from books and artwork to shoes and blankets—that enrich our daily lives. Manufactured capital can contribute directly to human well-being by providing food products, shelter, safety, and comfort; and it is essential for efficient access to natural capital. With the majority of people worldwide now living in cities, manufactured capital is ever more evident in the infrastructure of these urban systems and their connections to resources in other areas. Manufactured capital often plays an active part in people's health—for example, think of the role of the sewage systems in first spreading and later preventing disease outbreaks in London, or the role of London's crowded wooden housing in the Great Fire that scoured the city in the seventeenth century.

The amount of manufactured capital in place around the world has been accumulating rapidly. The accelerating rates of urbanization and consumption discussed earlier suggest that such growth is almost certain to continue, though at a decelerating rate in some of the regions around the world that are already highly industrialized and

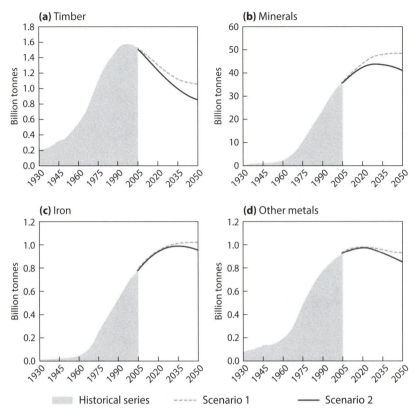

FIGURE 2.5. Change in estimated stocks of materials (timber, minerals, iron, and other metals) in constructed infrastructure in Japan from 1930 to 2005, as well as predicted trends, 2005–2030, under two scenarios. Scenario 1 predicts the future inflow of construction materials based on the average of the years 1995–2005. Scenario 2 considers Japan's future population, which is projected to decrease. Note the different scales. (Fishman, T., H. Schandl, H. Tanikawa, P. Walker, and F. Krausmann. 2014. "Accounting for the Material Stock of Nations." *Journal of Industrial Ecology* 18[3]: 407–20, fig. 3. doi:10.1111/jiec.12114)

have experienced slowed, stalled, or even negative population growth (figure 2.5).

Despite providing many benefits to human well-being, manufactured capital can also cause damage to other capital assets and thus potentially undermine the pursuit of sustainability. For example, destruction of natural forests to provide resources for manufacturing can cause decline in a range of ecosystem services (including biodiversity and erosion control), leading to an overall decline in natural capital. Likewise, emissions of pollutants through the production of energy for manufactured capital can lead to loss of human health and damage to human and natural capital, with long-term consequences for well-being. Importantly, these negative relationships are not inevitable. Managed forests can be harvested sustainably, with little long-term decline in natural capital. Emissions from manufacturing can be pushed toward zero, and resources can be reused again and again, minimizing the long-term impact on natural and health capital. As we will discuss further, the development of energy resources and manufactured materials with few negative impacts on other capitals holds great promise for sustainability goals.

As with natural capital, the goods and services produced from manufactured capital often flow from less developed to more developed areas of the globe, while many of the negative impacts (pollution, poor working conditions, and wages) remain in less developed areas. This inequity is driven both by consumer patterns (the richest 20% of global citizens account for over 76% of total private consumption)[35] and by the more stringent environmental regulations in wealthier countries. One result is that residents of more developed regions may not realize the full costs of the manufactured capital they create or use, because those costs are not clearly visible in their own areas. We will return to this problem of "invisibility" in the discussion of complex social-environmental systems in chapter 3.

New and exciting work is now being done by scientists and engineers to design manufactured capital that can better promote the goals of sustainable development. Green design, dematerialization or decarbonization of manufacturing processes, expansion of biotechnologies, and closed-loop "cradle-to-cradle" designs for reducing waste are among the many new directions being fostered (for some examples, see box 2.3). Eco-industrial parks similarly try to create efficiencies by collocating companies with complementary needs (e.g., the waste from fish farming in Denmark is repackaged as fertilizer for nearby farms).[36] These types of innovations are necessary to reduce the trade-offs now experienced when the growth of useful manufactured capital degrades other capital assets.

BOX 2.3. Cradle-to-Cradle Design:
Transforming the Purpose of Manufacturing

Cradle-to-cradle design is revolutionary in that it seeks to create products and processes that eliminate waste and contribute to well-being.[37] The design is so named to contrast with traditional "cradle-to-grave" manufacturing, where the product eventually becomes waste in the environment. Cradle-to-cradle design seeks to mimic natural growth and decomposition cycles, in which the raw materials in our environment are endlessly recycled. Cradle-to-cradle design incorporates concepts like dematerialization (reducing the amounts of materials used to create products) and decarbonization (reducing reliance on fossil fuels for manufacturing) but seeks to go beyond these efforts to be even less harmful and to embrace the possibility of inventing processes and products that actually benefit both human and environmental health.

Two pioneers of the cradle-to-cradle design movement, architect William McDonough and chemist Michael Braungart, founded a company[38] to help corporations think differently about their products. In construction, for instance, they challenge people to imagine "buildings that make oxygen, sequester carbon, fix nitrogen, distill water . . . accrue solar energy as fuel, build soil, create microclimate, change with the seasons, and are beautiful."[39] This type of cradle-to-cradle thinking is driving changes across many manufacturing sectors. For instance, Shaw Industries, one of the largest manufacturers of carpets in the world, has created carpet tiles that can be continually recycled without any loss in material. Furniture manufacturer Herman Miller recently redesigned an office chair to be 96% recyclable and to eliminate toxic chemicals, and now has a corporate goal of creating cradle-to-cradle designs. Method cleaning products have more than sixty different formulas that are cradle-to-cradle certified. While not every product is likely to become entirely sustainable, any product can benefit from a cradle-to-cradle ambition.

From the implementation of well-understood resource approaches such as composting; to the development of new engineering approaches to harvest materials from wastewater streams; to the creation of new, fully degradable or reusable building materials; to other new biotech and information technology-based innovations, the world is entering an exciting time as society seeks to dramatically increase what can be done with the resources in hand while limiting impacts on those assets or other components of social-environmental systems.

Human Capital

We argued earlier that key constituents of well-being for many people include their health and educational attainment. At the same time, however, healthy and well-educated people are powerful means for the realization of well-being. Thus, people hold a dual position in our treatment of sustainable development. We believe it is important, nonetheless, to articulate the specific role of **human capital**. In doing so, we have found it helpful to focus on three components that together determine the quantity and quality of humans as a capital asset: (1) the size, age structure and geographic distribution of the human population; (2) the health of that population; and (3) the acquired capabilities (education, experience, tacit knowledge) of the people who constitute that population.[40]

Drawing on a body of scholarship from demography, public health, and development policy, scholars and practitioners are beginning to understand how the overall size and health of populations are affected by changes in manufactured capital and other capital assets (including food and water, air or water pollution, occupational opportunities and risks, condition of housing [including exposure to indoor pollutants and pathogens], and others). Understanding some of the barriers to improved well-being can help spur interventions that might best reduce such barriers. Likewise, knowledge about the relationship between education received by various populations and measures of their well-being can provide insights into avenues that will most aid sustainable development. Overall, it is helpful to think of optimal human capital as "healthy, well educated, skilled, innovative and creative people."[41] A key task of sustainable development is figuring out how to foster these characteristics.

The state of human capital today is highly varied around the globe. We reviewed the status of health and education in our discussion of well-being earlier in this chapter (see figures 2.2 and 2.3). In terms of population, the world in which we are pursuing sustainability is getting more crowded, more urban, and older. The total population passed the seven billion mark in 2011. The annual growth rate of the global human population is just over 1% per year, down from twice that in the mid-1960s. The global population growth rate is expected to continue to

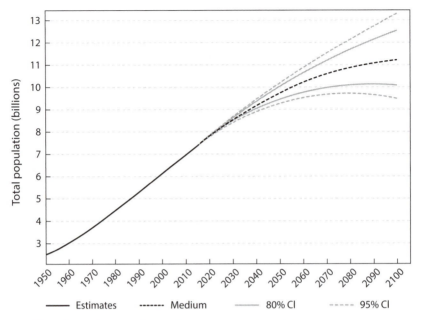

FIGURE 2.6A. Population of the world (in billions: 1950-2015 estimates based on data measurements; 2015-2100 according to the medium variant projection and confidence intervals (CI)). The medium variant projection "assumes a decline of fertility for countries where large families are still prevalent as well as a slight increase of fertility in several countries with fewer than two children per woman on average." (From World Population Prospects: The 2015 Revision, Key Findings and Advance Tables. Working Paper No. ESA/P/WP.241, by United Nations Department of Economic and Social Affairs, Population Division, © 2015 United Nations. Reprinted with the permission of the United Nations)

decline, reflecting falling total fertility rates (number of children a woman has in her lifetime) in most places around the world. These declining fertility rates are tightly related to access to health care, family planning, and women's access to education and employment opportunities; in other words, this trend of declining and, ultimately, zero or even negative population growth rates is one result of development. The UN and other sources estimate that the global population will reach nearly eleven billion by the end of this century, though this medium estimate could change dramatically as a result of changes in fertility (figure 2.6a). Most analysts expect that by the end of this century, or perhaps early in the next, the world's human population will begin to stabilize. Even if this trend continues, however, the number of peo-

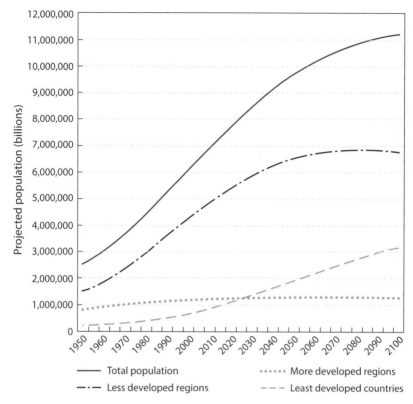

FIGURE 2.6B. Projected population (in billions) by development region, 1950–2100, according to the medium variant projection. "Less developed regions" include China and India but exclude least developed countries. "Least developed countries" include much of sub-Saharan Africa. (Created from data file "Total Population–Both Sexes," esa.un.org/unpd /wpp/DVD. United Nations, Department of Economic and Social Affairs, Population Division. 2015. *World Population Prospects: The 2015 Revision*; downloadable files)

ple *added* to the world population during the lifetime of today's young people is likely to exceed the world's *total* population in the middle of the twentieth century.

Almost all the future growth in population will happen in today's developing world—much of it in some of the poorest regions on earth (figure 2.6b). In contrast, many of today's most developed regions are expected to experience declining populations—a trend already under way in several of them. Increased migration will almost certainly be

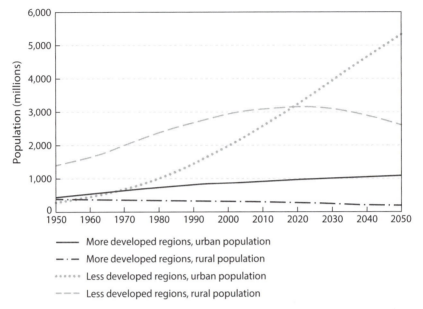

FIGURE 2.7. Urban and rural populations, by development group, 1950–2050. (From *World Urbanization Prospects: The 2007 Revision*, United Nations Department of Economic and Social Affairs, Population Division, © 2008 United Nations. Reprinted with the permission of the United Nations)

associated with these disparities, though other factors ranging from war to disease epidemics to environmental disasters to national legislation will also matter. That said, it is important to keep international migration rates in perspective: over recent history they have averaged only 10% or so of global birth rates.

Spatial redistribution of people within countries or regions is another matter. Urbanization has taken the world from the predominantly rural place it has been throughout human history, through a 50/50 split early in the twenty-first century, toward one that many forecast will be, on average, as urbanized as Europe is today (~70%). By the middle of this century, the vast majority of people living in both more and less developed countries will live in cities (figure 2.7). Part of this growth in urban populations will be due to an excess of local births over deaths, but a great deal will be due to rural–urban migration, much as it was in the historical London portrayed in our case study.

What do these population trends mean in terms of the relationship of human capital to the other capital assets and to well-being? Clearly, more people means more mouths to feed, with the potential (as we saw in the Nepal and Yaqui Valley cases) to damage natural capital in ways that affect human well-being. However, more people also means more hands to work (human capital) and more brains to discover and invent new ways to promote sustainability (knowledge capital). Concentrating that work in urban areas has historically opened the door for more efficient use of energy and building materials. Concentrating those brains in urban areas has historically been associated with accelerated rates of education and innovation. Concentrating all those people in urban areas has both intensified challenges to human health (whether by pollution or disease) but also increased the prospects for delivering better health care.

How these multiple dimensions of population growth and distribution work out in particular places depends in part on consumption choices and consequences. Policy and behavior changes (supported by social capital) may be needed to move higher-consuming countries toward sustainability at the same time that the poorest consume more. Social and knowledge capital will also become ever more important in reducing the impact of consumption—for example, through the production and use of less material-rich goods and services, more complete reuse and recycling, and diminished environmental impacts of consumption. The importance of social capital for a transition to sustainability is the topic of the next section.

Social Capital

The human capital embodied in individual people is important, but so is *social capital*—the connections among individuals and the norms of reciprocity and trustworthiness that arise from them. **Social capital** includes the economic, political, and social arrangements—including laws, rules, norms, networks, financial arrangements, institutions, and trust—that influence how people interact with one another, the environment, and other components of social-environmental systems. **Institutional arrangements**, as we use the term here, are a society's "rules of the game" that influence how people interact with one another and the rest of the social-environmental system. Those rules can be formal or informal. Examples of rules include policy, regulations, local norms and customs, contracts, and property-rights arrangements. Rules not only specify the rights and responsibilities for using and managing social-environmental systems but also stipulate whose responsibility it is to monitor and enforce rule compliance.

Together, the multiple arrangements that constitute social capital form the infrastructure within which actors and agents of change set agendas, hold governments and businesses accountable, and allocate goods and services. Social capital thus influences how effectively society can leverage the other capitals to produce inclusive well-being. For example, use of new fish tracking and harvesting technologies (manufactured capital, benefitting from knowledge capital, and supported by financial instruments and other forms of social capital) in marine fisheries can degrade or improve fish stocks (natural capital), depending on the presence and effectiveness of fishing policies and norms (social capital). In our London case, reducing the effects of water contamination on human health in the nineteenth century required not only new knowledge about the relation of contaminated water to disease (knowledge capital) and innovative technologies such as sewers (manufactured capital) but also the creation of new regional governance and financing structures (social capital) that allowed those improvements to be implemented at scale. Social capital plays a particularly important role with respect to access and use of natural capital. What types of power relationships exist, how markets and property rights are structured, and who has access to decision making are key to determining whether natural capital in a given area is degraded or not.

How well is society doing in managing its social capital assets in ways that could help the pursuit of sustainability? This turns out to be a complex question for which there are many partial answers but no consensus. As we discuss later, in chapter 4, "good governance" of social-environmental systems—a component of social capital—is essential for sustainability if for no other reason than that so many of the challenges facing sustainable development require people to work together if they are to be resolved. There also seems little question that over the last half-century society has made substantial progress in strengthening relevant rules and norms covering topics ranging from human rights to democratic decision making to environmental protec-

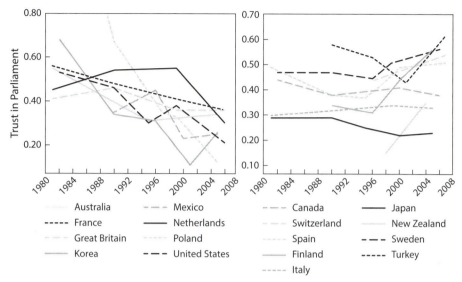

FIGURE 2.8. Trends in trust in parliament—which is considered here as the key representative institution of democracies—in selected OECD countries. The figure is based on data from the World Value Survey. The group of countries on the left show a pronounced decline in trust in parliament: these include Poland, Korea, the United States, as well as Mexico, France, and the Netherlands. The group of countries on the right—in particular, Sweden, Turkey, New Zealand, and Spain—show increasing or constant levels of trust in parliament. (Morrone, A., N. Tontoranelli, G. Ranuzzi. 2009. "How Good Is Trust?: Measuring Trust and Its Role for the Progress of Societies." *OECD Statistics Working Papers* 3: 1–38, fig. 4. OECD Publishing. doi:10.1787/220633873086)

tion. Nonetheless, a critical look at the evidence suggests that these efforts are falling short of what is needed in multiple ways. Global environmental treaties have largely failed, the important exception described in our case study on protection of the ozone layer notwithstanding. The rise of neoliberal deference to markets, and hostility to government, has weakened social resilience. And the degradation of the multiple interdependent ties that help communities unite in common purpose is being noticed and lamented around the world. One metric widely accepted to be central to all these manifestations of social capital is trust. But trust—whether in governments or in fellow citizens—is spottily distributed around the world, and in many places has been degrading for decades (figure 2.8).

Improving social capital worldwide is very important in terms of navigating paths to sustainable use of natural capital stocks. It is also essential for reducing human vulnerability (defined as the likelihood of suffering harm) to social and environmental shocks. For example, in most situations around the world the poorest and most disadvantaged are currently also the most vulnerable to climate-related events. People living in city slums around the world, without access to public services or resilient infrastructure, are especially vulnerable to storms, droughts, heat waves, and other climate events. Likewise, countries with low social capital (e.g., with unstable political systems that allow continuing violent conflicts) are not likely to put in place long-term credible incentives to reduce environmental effects of development. Thus, researchers are focusing on institutional change and other social capital enhancements to assist in both individual and collective adaptation to climate change.[42]

Scholars in the social sciences as well as practitioners who are interested in sustainable development are also asking broader questions such as the following about how people develop social capital to promote prosperity: How can resources be allocated efficiently? How can innovation for sustainability goals be stimulated and financed? How can corporations succeed financially while including concerns for future generations? How can the political will and trust needed to collaborate in the development of shared resources be built? Encompassing these concerns is the more fundamental question of how society can promote a fair sharing of the benefits and risks arising from their use of capital assets. In other words, how can institutions be strengthened to avoid the "tragedy of the commons" that threatens when too many demands are placed on shared limited resources? We discuss some of the answers to these questions at greater length in chapter 4, which explores how governance processes in social-environmental systems affect efforts to further the goals of sustainable development.

Knowledge Capital

Finally, **knowledge capital** is the fifth asset stock on which society depends in efforts to enhance its well-being. As used in our framework, knowledge capital encompasses both conceptual and practical knowledge, including general principles, information, facts, device designs, and procedures. These are all intangible public goods: they can in principle be used by anyone who wishes to do so, and they can be used repeatedly, without depletion.

Some portion of the stock of useful knowledge is always declining, whether through technological obsolescence (as did the skills involved in sailing three-masted warships) or as a result of biological evolution (as in the loss of efficacy of overused antibiotics or pesticides). Research and innovation therefore constitute essential investments in resupplying, renewing, and expanding the stock of public knowledge. Likewise, the assembly for new uses of previously created knowledge and the integration of tacit, private, or experiential knowledge into the knowledge bank expand the stock of public knowledge. However it is created, knowledge is inevitably incomplete and fallible in the face of the complex realities of sustainable development. Thus, adaptive approaches are needed that can rapidly feed useful information from field experience back into evaluation and redesign of innovations. The design of such adaptive approaches to enhance the stock of useful knowledge is an area of considerable research.

Knowledge capital cannot realize its full potential to promote sustainable development unless it is supported by other capital assets. Strong social capital is needed to incentivize and make accessible innovation, particularly in the area of public goods. Also necessary is human capital that is sufficiently well educated to utilize knowledge once it exists. And manufactured capital such as information technology can play an important role in assuring that knowledge is equitably accessible to a broad cross section of society. When society's management of its capital assets succeeds in promoting such synergies, the contribution of new knowledge to sustainable development can be profound. Indeed, many observers have argued that the pursuit of sustainability must have at its core a capacity to effectively promote adaptive innovation as well as broad access to the fruits of such innovation.

Many people around the world are investigating how knowledge for sustainable development can be created and mobilized so that it improves well-being inclusively—for today's underprivileged populations, for those who are now well off but want to reduce the deleterious "footprints" of their consumption, and for generations yet unborn. In chapter 5 we discuss their successes, failures, and lessons learned in creating (and assuring equitable access to) the knowledge capital needed for pursuing sustainability.

TOWARD THE INTEGRATION OF CAPITAL ASSETS FOR INCLUSIVE WELL-BEING

It should be clear from the preceding discussion that the five capital asset groups can interact in ways that improve human well-being; indeed, all the improvements in human well-being that we discussed at the beginning of this chapter can be traced to the joint utilization of these assets. Looking to the past, it is also clear that many improvements in social well-being have been made at the particular cost of natural capital: food production has been increased by clearing grasslands and forests and irrigating deserts; sickness has been reduced by draining swamps that harbored disease; light, heat and power have enriched the lives of billions through the mobilization of fossil fuels with their attendant pollutant emissions. The benefits have been great, but many of them run the risk of being short term, because the costs in terms of natural capital are too high.

So what is to be done? How can society strive to meet the needs of people today while also protecting the natural capital and other assets that make up our planet's life support systems? How can capital assets be managed to achieve improvements in well-being that are truly inclusive: extending around the world and lasting across generations?

We believe it is urgent that these questions be answered, and finding answers presents enormous opportunities. People are learning how to increase consumption of what they truly need while reducing consumption of what they don't. Through deployment of new management approaches and innovative design of new technologies, the negative impacts of production and consumption are being reduced. Means of improving current well-being are taking into account the need to sustain the diverse capital assets on which the well-being of future generations depends. New kinds of social capital—including social and environmental policies in corporations and governments, international agreements, and local knowledge networks—are encouraging and regulating the creation of manufactured capital that benefits people while protecting the environment. Healthier and more educated populations are more able to conceive and implement new ways of doing business on the planet. New kinds of knowledge and its integration with experience are leading to better ways of moving forward. For a transition to sustainability, all these pieces need to fit together. We believe our framework for thinking about sustainable development encourages us to focus on the integrated management of all the capital assets to shape improvements in well-being now and in the future.

Fitting the pieces together, however, requires an ability to understand how changes in one capital asset—whether brought about by changes in policy, taste, or technology—are likely to interact with the other assets and thus to work their way through the processes of production and consumption to shape inclusive social well-being. Doing this perfectly is not within our reach. But the revolutionary advances of the last decades in understanding the dynamics of social-environmental systems provide a solid foundation on which to do it better. Making sense of what that understanding has to offer for pursuing sustainability is the focus of our next chapter.

CHAPTER 3

Dynamics of Social-Environmental Systems

Social well-being rests on many factors. As we saw in chapter 2, education and employment opportunities; technologies and manufactured goods; well-functioning governments, corporations, and institutions to support the workings of society; and the goods and services provided by the planet's natural capital—humanity's "life support systems"—all matter. For sustainable development, they all have to work together— they are part of a tightly coupled system.

In the framework for understanding sustainability introduced in figure 2.1, we followed convention in portraying that tightly coupled system in terms of its production (supply) and consumption (demand) processes. This perspective is useful for analyzing how particular goods and services (e.g., food, energy, housing, etc.) can be produced and consumed in a manner more consistent with sustainable development. If we were writing a text centered on one group of those goods and services (e.g., on the role of the food system in sustainable development), we would almost certainly stick with the production-consumption framing. As we suggested in chapter 2, however, one drawback of that framing is that production and consumption processes are seldom as distinct as they seem when set up in opposition to each other. Rather, each is a cause and a consequence of the other. Another disadvantage is that a focus on particular production-consumption systems tends to obscure the competing demands of other such systems on the capital asset base (e.g., food and energy systems competing for the same water). Finally, the production-consumption framing implies that human choices are all that matter, downplaying the role of nature's own dynamics (e.g., evolution, natural climate change). For all these reasons, we have found it important in the pursuit of sustainability to look beyond particular production-consumption systems to the general properties of the social-environmental systems that set the stage for sustainable development. Those properties, and the dynamics that flow from them, are the topic of this chapter.[1]

Understanding how social-environmental systems work is needed to improve predictions of how specific interventions (e.g., new policies, new technologies, new management approaches) are likely to add to or detract from social well-being. However, most scientists, managers, and decision makers have tended to focus on one or another part of social-environmental systems, with little attention to their coupled nature. Development organizations and social scientists have most often focused on human, social, and economic issues, activities, and interactions, setting environmental concerns aside and considering resource issues only in the near term. Conversely, conservation groups and natural scientists in the past have often focused on and given precedence to the environment and "natural world," with people considered largely as negative pressures that put ecosystems and environmental systems at risk. To meet the goals of sustainable development, the knowledge gained from these narrow perspectives has to be merged. Without an integrated appreciation and understanding of the social-environmental systems in which decisions are being made, unintended and negative consequences will too often result.

The cases introduced in chapter 1 provide some good illustrations of well-intentioned approaches that while succeeding in improving some aspects of human well-being have fallen short in their long-term contribution to inclusive social well-being. In our case study about the Yaqui Valley, the home of the Green Revolution for wheat, the dramatic increase in food production between the 1950s and 2000s rested on the intensification of agriculture through the use of improved cereal varieties, industrial fertilizers, and irrigation. The Green Revolution was all about meeting the needs of a growing human population for food— a very important, well-intentioned objective! Unfortunately, that intensification was sometimes accompanied by leakage of fertilizer and pesticides with resulting negative effects on air and water and human health; overuse of water resources; degradation of soil resources; social inequities; and displacement of communities and indigenous groups. Today's challenge in the Yaqui Valley and elsewhere around the world is to continue to increase food production on land already under agriculture while reducing the negative consequences that will undermine the well-being of future generations.

Similarly, the problem with stratospheric ozone started with an effort by scientists to improve the well-being of people. The search for safe refrigerants led to the invention of a wholly new class of chemicals, called CFCs, that seemed safe as well as useful for many other critical manufacturing processes and products. The story up to that point was a great example of science and engineering innovations focused on social challenges and human well-being. Scientists working for the manufacturing companies conducted extensive tests of CFCs to make sure they raised no environmental concerns, and gave them a clean bill of health. Unfortunately, however, they and everyone else missed (because the science of the day lacked a system perspective and didn't consider or measure what happens to CFCs high in the atmosphere) the fact that CFCs could destroy a critical protective layer high in the atmosphere, with potentially negative outcomes for human and ecosystem health. The emerging knowledge of that potential led, over time, to an international agreement that limited the use of CFCs and other ozone-depleting substances. Loss of ozone was no longer considered an "environmental problem" but was acted on because it threatened human health and well-being. (That is, the framing of the problem shifted from "Save the ozone layer!" to "Save us!")

In the long history of London's attempts to manage the needs of a growing city, misunderstandings about social-environmental systems led to human suffering and environmental damage. For example, Londoners of the early nineteenth century did not understand that cholera was caused by bacteria that could be spread through contaminated water supplies. Thinking such diseases were caused by foul odors, the city expended major efforts to drain human waste out of backyards and cesspools into the Thames River. Given that the Thames was a central source of water for residents, this effort in trying to clean the city's environment inadvertently led to the death of thousands of residents from cholera and other diseases. Eventually, through sustained, collaborative, and interdisciplinary work by scientists and citizens, the source of disease was discovered and action followed.

In the case of irrigation in Nepal, consider what happened when government interventions in irrigation did not take the engagement and concerns of small social communities into consideration. Clearly, not recognizing or, worse yet, ignoring the linkages and interactions in social-environmental systems can cause problems.

How, then, should we think about these systems? What do we know about coupled social-environmental systems and how they can be analyzed and managed?

UNDERSTANDING SYSTEMS

The recognition that humans, their institutions and infrastructure, and their environmental life support systems are part of a coupled system leads those of us pursuing sustainability to ask, how do those systems work? and how do policies, management practices, and other interventions affect the workings of the system? Social-environmental systems are **complex adaptive systems**—systems that have multiple interconnected components that interact in diverse ways. They exhibit positive and negative feedbacks, connections across space and time, and nonlinearities and tipping points that influence the way the system works and the way it changes with each new intervention. But before diving into some of what is known about social-environmental systems—the topic of this chapter—let's take a brief look at characteristics of systems in general.

Whether used by an engineer describing a car or a wastewater treatment plant, by an ecologist describing a forest, or by an urban planner describing a city, the term **system** generally means a bounded area with a set of elements or components that are connected to and interact with each other. A fisherman and his boat can be considered a system for fishing. But putting them on a lake or ocean creates a new system that now includes water, fish, the food they eat, the predators that eat them, the other organisms that compete with them, the range of fisher communities and the gear that harvest them, and a set of regulations to govern the fisherman's actions. To analyze and explain what's happening in and to manage a particular system, one needs to define the system. In other words, the boundaries of the system have to be described.

Social-environmental systems can be defined, studied, and evaluated at many spatial and temporal scales. Our case study of London, for example, focused on an "urban" scale of tens of kilometers but looked "down" from there to the scale of households and neighborhoods and "up" to the metropolitan area and larger hinterland from which the city drew its capital assets. Our ozone case, in contrast, focused on the global scale, from the surface of the planet to the outer atmosphere and all around the planet. Our Nepali irrigation case focused on the watershed scale but, again, needed to consider both household and national phenomena. In the Yaqui Valley case, scales ranged from the field and farm scale to the agricultural district to the ocean-land regional scale to the state, national, and global scales (figure 3.1). Moreover, connections that link across and between each of these scales are important. In the Yaqui Valley case, for example, decisions made at the field and farm scale made great sense there but were connected—with sometimes unfortunate consequences—to other places and scales through the transfer of water, air, people, other species, and money, among other things. Thus, farm-scale actions led to unintentional effects on downstream water quality in streams, aquifers, and ocean; and upstream air quality, including in populated urban areas of the valley, as well as at a global scale.

Time scale is also important in defining systems of interest. One of the reasons we included the London case in this book is to emphasize the importance, and the difficulty, of seriously grappling with the long time scales inherent in the intergenerational concerns of sustainability analysis. Given the multiple catastrophes that have befallen London over its two thousand–year history, what does it mean today to say that London's development has (or has not) been sustainable over that period? Looking to the future, should an evaluation of London's prospects for sustainable development try to look ahead a generation, a millennium, or another period? What time frame should a sustainability analysis of London have considered had it taken place in 1670, in the immediate wake of the Great Plague and Great Fire and with one of its leading citizens bemoaning that "the City [was] less and less likely to

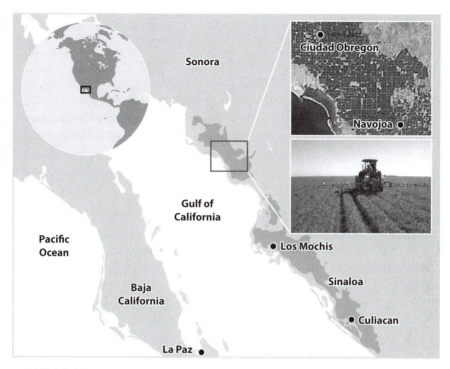

FIGURE 3.1. Various scales in the Yaqui Valley. (Farm field ≈ 20–50 hectares; agricultural region: 233,000 hectares; 100 hectares equals 1 square kilometer.)

be built again, every body settling elsewhere, and nobody encouraged to trade"?[2]

What is known is that the functioning of systems is very much influenced by what happened to and within it during its earlier history. Economic development has always had a sense of history and temporal dynamics. The study of ecosystems was dramatically changed when ecologists began to realize that ecosystems are not often in "equilibrium" but that past events (like land clearing or hurricane disturbances) influence how the system functions today and will continue to do so in the future.

Spatial and temporal changes in systems can be understood and tracked by measuring the changes in stocks (or amounts, such as the amount of biomass in the trees of a forest, the amount of water in a storage tank or aquifer, the number of people in an office building or city, or the stocks of capital assets introduced in chapter 2) of particular components of the system over space or time. In any system, the size or amount of the stock is controlled by inputs and outputs (or inflows or outflows). In a classic example, the amount of water in a bathtub at a particular point in time (the stock of water), and its change over time, is a function of how much water is flowing in minus how much water is flowing out. Opening or closing the faucet influences the amount of water coming in, and opening or closing the drain influences the amount of water going out; together, these inputs and outputs control the stock of water. For Nepali farmers hoping to control water flows into and out of their fields, this becomes a very real exercise—how large the stock of water is relative to inputs and outflows can make the difference between a failed harvest and a successful one. In some systems, the size of the stock is very large compared with the inputs or outputs or their difference. In these systems, there may be a long time lag before the effect of changing the controls is seen. Thus, there is said to be inertia in the system. Inertia in critical stocks sometimes means that "the ship can't turn on a dime," and planning is needed (as is the case with climate change).

UNDERSTANDING COMPLEX SOCIAL-ENVIRONMENTAL SYSTEMS

With these characteristics of all systems in mind, how can social-environmental systems be understood and evaluated? In such systems, the five capital assets that we described in chapter 2 (natural, manu-factured, social, human, and knowledge capitals) are the relevant "stocks." The flows into and out of those asset stocks are controlled by many natural processes but also by society's processes of production and consumption. As we suggested in chapter 2, sustainable development is in many ways a challenge of managing assets to promote in-clusive social well-being. Before a course of action is set or an inter-vention made, it therefore behooves us to evaluate how those actions are likely to change the flows into and out of the asset base of the social-environmental system, and thus the size and composition of the asset stocks and their capability to support sustainable development.

The ability to understand the consequences of different production and consumption decisions for the state of the asset base seems essential for the informed pursuit of sustainability. Unfortunately, however, complete understanding of those consequences is seldom within reach. Why is this the case? It is because social-environmental systems are complex adaptive systems.

Complex adaptive systems are widely studied in ecology, biology, physics, mathematics, computer science, anthropology, and planning. As we noted earlier, the term denotes systems that have multiple interconnected components with feedbacks affecting their interactions, and that are characterized by self-organization and emergent behavior (in which the system as a whole is more complex and organized than would be predicted by the behavior of the individual parts). The following sections discuss some of the most challenging features associated with complex social-environmental systems—factors that make the informed pursuit of sustainable development difficult indeed.

Feedback Interactions in the System

One of the most important ways that elements of systems interact is through **feedback loops**. Feedbacks result when change in one part or component of a system—a particular process or variable—affects other parts in such a way as to loop back to influence that original component, either by reinforcing the change in that component (a positive feedback) or dampening the change (a negative feedback). Feedbacks result from a chain of events that can sometimes be very hard to measure or predict. Let's look at a few examples of feedbacks that are relevant to sustainable development.

In underprivileged communities lacking in adequate public services, efforts have been undertaken to improve infrastructure and services—through the building of roads and schools and playgrounds and safe water infrastructure, perhaps. These efforts are designed and undertaken purposefully to improve the quality of living for the residents of the community; all else being equal, they should do so. However, a feedback is possible: with better infrastructure and support, the area becomes a draw for people in need, leading to a re-creation and possibly worsening situation of inadequate public services. Only ongoing investment to improve infrastructure and services can save the place from declining more.

Likewise, the presence of a pristine or high-quality ecosystem (such as a coral reef surrounding a remote tropical island) can positively affect local and national economic well-being through tourism, but overdevelopment of the tourism industry and resulting overexploitation of the ecosystems goods and services can make those benefits very short-term, eventually feeding back to degrade the ecosystem and reduce tourism benefits—and ultimately reducing economic well-being. Similarly, new fishing technologies (including the use of satellite tracking of fish stocks) put in place by commercial fisheries have been incredibly successful in improving catch per fishing effort, thus temporarily improving fishers' economic well-being, but at the same time have contributed to dramatic levels of overfishing that have caused some fish stocks to crash and have ultimately put some fishermen and fisherwomen out of business. Clearly, feedbacks can lead to unintended consequences if they are not anticipated and managed.

Our case studies, not surprisingly, illustrate a number of feedback mechanisms. The most severe ones are evident in London's record of population growth (appendix A, figure A.1): periods of rapid growth were invariably followed by long periods of little change, each reflecting feedbacks from overstressing one resource or another that decreased the viability of the city as a place to prosper. In the Nepal case, successful farmer-managed irrigation systems constructed a crucial feedback to address the "free riding" that was lacking in the failed government managed systems. The farmers did this by negotiating among themselves clear rules for what each household needed to contribute and put in place an agreed-upon system for how members of the community would monitor and enforce compliance. In the ozone story, some feedbacks can be found in atmospheric chemistry (recall the chain reaction that leads to ozone breakdown), but it also involves a fascinating example of a social feedback mechanism. The Montreal Protocol was successful in part because its agreements built in the possibility that new information—from monitoring, scientific studies, or technological advances—could be used to adjust the timeline for reductions in use of ozone-depleting chemicals. Such feedback proved invaluable, resulting in a series of updates to the protocol that repeatedly strengthened it as new, increasingly worrisome evidence became available.

Invisibilities in Space and Time

The job of making smart choices that promote inclusive social well-being is further complicated by the fact that the parts of social-environmental systems are linked with each other across space and time. Decisions made in one place or time can have consequences in places and times far distant. When those consequences are "visible"—known, understood, and believed—to all concerned, there is at least some prospect that governance agreements can be negotiated to assure that one group's well-being is not advanced at the expense of others. But what if local choices affect other communities or generations in ways that the decision makers can't see? Such **invisibilities** abound in social-environmental systems and are often discussed as **externalities** by economists and decision analysts. We'll come back to the governance of externalities in chapter 4. Here, we focus on the role of science in making them visible to decision makers and concerned citizens so that governance is possible.

The first dimension of invisibility is simply ignorance. When the natural capital asset coal was first exploited to fuel the production of energy services like heat and light in twelfth-century London, no one knew that some of the pollutants emitted through its use could cause dangerous climate change. It would take 700 years for science to mature to a state in which the greenhouse effect was well understood, and another century for that science—and the monitoring it led to—to advance to the point that it made the risks of climate change resulting from fossil fuel use sufficiently "visible" to society that its leaders began to address the problem. The ozone issue moved faster, but when CFCs were first synthesized and added to the world's stock of knowledge capital in the 1930s, the state of scientific understanding was such that no one could have any idea that this discovery posed risks to the stratospheric ozone layer and thus to human and ecosystem health. Forty years would pass before research and monitoring efforts—both pursued for reasons largely unrelated to CFCs—would make those risks sufficiently visible to begin stirring political action.

Invisibilities caused by ignorance are ubiquitous in social-environmental systems. Some are made visible as a result of targeted search, such as the quest of epidemiologists to discover whether particular chemicals in the environment pose a risk to health. At least as important for the pursuit of sustainability are curiosity-driven research and monitoring that cast light into the dark spaces of our ignorance, thereby making visible the new dangers as well as the opportunities hidden there.

A second sort of invisibility results when decision makers in one place can't see the damages (or benefits) of their actions to people who live someplace else. Since the overall social consequences of their actions are invisible to them, decision makers are therefore likely to make "selfish" choices that may undermine inclusive social well-being and thus sustainable development. The Nepal irrigation case provides a good example of problems that arise when upstream water users do not see, and cannot be held accountable for, the downstream damages resulting from their decisions to withdraw more than their share.

At a larger scale, transboundary air pollution poses similar challenges. Consider, for example, efforts to deal with this problem in Europe. As shown in our London case, the deleterious local effects of air pollution resulting from the use of fossil fuels were visible to and acknowledged by people there as early as the thirteenth century. Over time, decision makers in London (and Great Britain more broadly) mitigated these unwelcome effects in a number of ways, including by switching fuels and improving technology but also by erecting increasingly taller smokestacks that sent much of the noxious pollution far away: out of sight, out of mind. It was not until the mid-1950s, however, that Swedish scientists began to make visible where that pollution was going, that is, across national borders to be deposited over the rest of Europe. Thirty years of intensive research and monitoring was required, however, before Great Britain's responsibility for significant air pollution damage to others in Europe became irrefutably visible to its public and political leaders. Faced with this evidence, Great Britain agreed to join other countries in signing an international agreement to reduce the pollution burden that each imposed on the others through its own consumption of energy derived from fossil fuels.

A third dimension of invisibility arises when decision makers in one place can't see the damages (or benefits) of their actions to future generations. Once again, since the overall social consequences of their actions are invisible to them, decision makers are therefore likely to make "selfish" choices that may undermine inclusive social well-being and thus sustainable development. Many examples exist, ranging from the depletion by agricultural production practices of natural capital essential to future generations (e.g., soil and groundwater) to the buildup of slow-acting carcinogens in our water supply. The classic case, however, is again that of greenhouse gases and climate change.

Contemporary science, such as that reported by the Intergovernmental Panel on Climate Change,[3] paints a pretty devastating picture of what the future climate could look like if present practices of energy production are not radically changed. Yet, for many political leaders (and the interest groups that fund them), today's benefits from continuing to use fossil fuels are far more visible and compelling than the dim and distant specter of a sizzling and sunken future. Many sectors of society—including social activists, novelists, and filmmakers—are working hard to make more visible the scientific knowledge about the likely future consequences of a continuing carbon-based energy economy, and increasingly, the effects themselves are becoming visible.

Making the invisible visible is an essential task in the pursuit of sustainability. Essential contributors to this process are scientific research, analysis, measurements, and monitoring. It is therefore not surprising that those who would prefer to realize their immediate self-interests while keeping the consequences of their selfishness invisible to people in other places and times also try to undermine science at every turn. Pursuing sustainability is not for the faint of heart.

Complexity

The most important implication of the complexity of social-environmental systems is that "You can't do just one thing." The phrase is attributed to many people, but its meaning is clear: feedbacks, invisibilities, and the sheer number of interacting components in social-environmental systems mean that any intervention is likely to have consequences well beyond its intended target.[4] In our case studies, we see multiple examples. The new "water closets" of early nineteenth-century London achieved their purpose of ridding houses and their adjoining alleys of foul-smelling human wastes. But by conveying those untreated wastes into Thames River, they inadvertently poisoned the city's principal source for drinking water. The innovation of CFCs greatly enhanced society's ability to provide safe refrigeration of food, but through a perversely complex chain of unforeseen connections it also put the world at risk by causing depletion of the stratospheric ozone layer. With the introduction of modern technology into the head-works of Nepal's irrigation system, it indeed worked better at controlling water. However, because the system was under the control of new technology and nonlocal managers, the local farmers lost their incentive to cooperate with one another, which led to a decay of overall system productivity. Numerous additional cases can be drawn from today's headlines—for example, the unintended effects on food prices of government subsidies to promote biofuels over fossil fuels.

The complexity of social-environmental systems means that all the consequences of new policies or technologies introduced in pursuit of sustainability cannot be predicted with confidence. Better science can help, but it needs to be science imbued with humility. The pursuit of sustainability in complex social-environmental systems needs to be an adaptive process in which the best possible interventions are tried, the results carefully monitored, and the course corrected when things don't go as planned.

Tipping Points, Regime Shifts, and Surprises

Even a careful adaptive approach with excellent monitoring has its limitations, however. Drive off a foggy road in the prairies, and a bit of adaptive backing up lets you continue on your way. Driving off a foggy road in the mountains is a very different situation indeed.

Social-environmental systems, like all complex systems, are more like mountains than prairies, characterized by nonlinearities and thresholds beyond which different relationships hold.[5] A broad and growing literature suggests that social-environmental systems can cross thresholds or **tipping points** at which even a small perturbation can alter the state and functioning of the system, leading to regime shifts. **Regime shifts** are large, persistent, and often abrupt changes in the dynamics of a system that occur because of substantial changes in interactions and

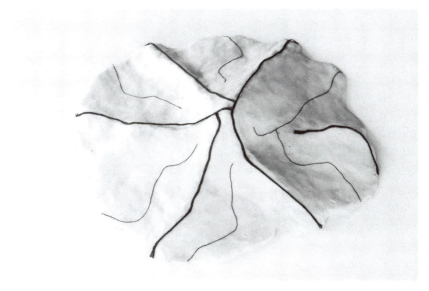

FIGURE 3.2. The regimes in this figure are the six drainage basins, each (let us assume) emptying into a different sea. The "critical thresholds" or "tipping points" are the ridges or divides separating the basins. A drop of water on a ridgeline (an "unstable equilibrium") has two or three possible trajectories and destinations awaiting it, depending on which basin ("domain of attraction") it happens to enter. But as long as the raindrop is still near the crest of the divide, a small "push" can move it back across the divide into the neighboring drainage with consequences for where the drop ends up. In typical landscapes, as the drop moves farther down a particular drainage, most "pushes" won't be large enough to move it over the increasingly higher ridges (more distant "tipping points") separating it from neighboring valleys. (In systems speak, its basic trajectory, though not its exact position, is "stable" to a wide range of perturbations.) Of course, a sufficiently large perturbation, whether from a strong wind or a particularly willful water drop, can lift the drop over even a very high divide and "tip" it into the neighboring drainage. Equally important, other forces can progressively erode or lower the divide ("tipping point"), making it easier and easier for the droplet to cross into a neighboring drainage basin and a new trajectory (regime). (© St. Johns River Water Management District)

forcings in the system. Such shifts are not readily seen until they are upon us, and they can cause quick and surprising transitions from one state or regime to another, with the new one functioning quite differently from its predecessor.[6] Figure 3.2 illustrates critical thresholds or tipping points that link the different regimes or domains.

Regime shifts and tipping points are hot topics of study in some areas of biophysical science. For example, Stephen Carpenter and his colleagues showed how gradual changes in nutrient enrichment slowly influence biological processes in lakes, but with very little apparent change in the way the lakes look. At some point, however—perhaps after a heavy rainstorm that washes large amounts of fertilizer phosphorus to the lake—the system reaches a threshold and tips into a new regime—from a lake with crystal clear water to one that is turbid and green with algae—with implications for drinking water quality as well as recreational use and other ecosystem services.[7] Recent carbon and climate-related changes in coral reefs, too, provide an archetypical example of regime shift—from hard corals to algae and soft corals—that has been extensively studied and mathematically modeled, and that has serious implications for human well-being. Many other tipping points and regime changes have been observed or hypothesized to be possible as a result of climate change. Some, such as the collapse of the West Antarctic ice sheet, could have serious implications for social-environmental systems worldwide and for intergenerational social well-being.[8]

Tipping points and the potential regime changes that lie beyond them are also key to understanding social aspects of systems. For example, they are essential to understanding how so-called poverty traps function and how they might be surmounted. In a poverty trap, poverty in a social-environmental system persists over time, "trapping" people and their descendants in a seemingly endless state of poverty. (Imagine a crater at the top of the mountain in figure 3.2. The "trap" would be water stuck in the crater, unable to get over its walls and thereby take advantage of the multiple development paths represented by the descending valleys.) Despite a variety of political and economic interventions, some areas of long-term poverty persist both within developed countries (e.g., the rural Appalachian region in the United States) and on a more global scale (e.g., within many sub-Saharan African nations). Much effort has been focused on finding effective ways to help people cross tipping points from regimes of persistent poverty into more dynamic trajectories of improving well-being. There is growing evidence of "critical asset thresholds" that people must cross to escape poverty.[9] Though the exact makeup of such thresholds will vary among places, this means that people need certain combinations of assets—human, knowledge, social, manufactured, natural—before their own hard work can help them get ahead. Significantly, natural capital deficits and degradation play a central role in many of today's deepest poverty traps, a fact that emphasizes the importance of comprehensive asset-based perspectives on sustainable development efforts such as the one we have presented here.

Vulnerability and Resilience of Social-Environmental Systems

Social-environmental systems are exposed to many different kinds of perturbations or external forces—stresses or disturbances or intentional interventions—that ultimately can make them function or work differently and can even lead to regime change. Changes in temperature and precipitation and storminess; land degradation and changes in soil quality; alternations in the numbers and types of plant, animal, and microbial species; policies that affect economic or financial opportunities; new technologies; and many others can all perturb how social-environmental systems function, with good or bad results for sustainability. Ultimately, perturbations interact with the dynamic characteristics of complex systems that we discussed earlier: feedbacks, invisibilities, tipping points, and regime shifts. The following are several possible results for the overall social-environmental system and for humanity's efforts to manage the assets of those systems in pursuit of sustainability:

- The system can absorb the perturbation and make some adjustments but continue much as before. (Our case study shows London doing this throughout most of its history.)
- It can collapse into its component parts and reassemble its core capacities very slowly if at all. (Following the double whammy of the Great Plague and Great Fire in the late 1660s, London grew hardly at all for a century and was characterized by contemporaries as "less and less likely to be built again, every body settling elsewhere, and nobody encouraged to trade.")
- It can transform the shock into an opportunity to do things in new ways that allow it to prosper as never before. (London responded to

the Great Stink and cholera epidemics of the mid-nineteenth century with innovations in finance, governance, and infrastructure that helped to launch a doubling of its population and its emergence, by the end of that century, as the greatest city in the world.)

What determines the capacity of social-environmental systems to cope with shocks and surprises? While the answer to that question is far from clear, it is the subject of a very active area of sustainability science that focuses on the themes of vulnerability, resilience, and adaptive management. **Resilience** is the ability of the social-environmental system faced with stresses, shocks, or surprises to continue to perform its current functions or even to extract benefits from those disturbances. **Vulnerability** can be defined as the likelihood of suffering harm. In the context of sustainable development, vulnerability is particularly useful as a lens through which to identify particular social-environmental systems and their component parts—people, infrastructure, ecosystems, resources, economic and governance systems, and so on—that are most likely to suffer damage when stresses, disturbances, and surprises occur. Even when the overall system is coping pretty well, a vulnerability perspective helps us see who or what is left out and needs particular attention if inclusive measures of well-being are not to decline.

How vulnerable a system or its parts happen to be depends on exposure to stresses but, just as important, also depends on access to capital assets: who can and who can't draw on which assets in an effort to cope? In our Yaqui Valley case, for example, the agricultural community managed to grow at least some crops even during a prolonged drought because it had large surface reservoirs as well as groundwater resources to partially tide it over, and government support to fill in financial gaps. That social-environmental system was less vulnerable to climate-related stresses than others that lack natural capital in the form of groundwater resources, or social capital in the form of government policies that support agriculture. Access to institutional structures and networks, education, insurance, and/or new knowledge and technologies for more efficient water use could also reduce vulnerability. More generally, it is largely differential lack of access to assets that make the poor, children, and the elderly more vulnerable in the face of acute disasters like floods and chronic ones like climate change.

What makes a social-environmental system resilient? While there is no definitive set of characteristics, there's a short list of features supported by many scientists and much experience in this field.[10] Among features commonly discussed are diversity, redundancy, and connectivity, all referring to both the social and environmental components of the system. On the environmental side, diversity encompasses biological diversity of species and genes and ecosystems and landscapes, but it can also include diversity in available resources (e.g., both groundwater and surface water). On the social side, diversity includes variety in cultural groups, livelihood strategies, management arrangements, centers of innovation, and other institutional forms (e.g., building and health codes or poverty alleviation measures). Redundancy or replication of some elements is also important to resiliency in that it provides some insurance against loss of others.

Examples of the importance of diversity in social-environmental system are broad ranging. Agroforestry systems, for example, are valued in many places because they have a range of crop and tree species that are harvested discontinuously over time so that a particular pest or drought or other stress does not lead to failure of the entire food production system. Likewise, diversity in types of organizations (e.g., government, non-governmental, community based, private) provides for some substitutability of support for social-environmental systems under economic, social, or environmental stress.

Connectivity denotes the ways in which resources, species, decision makers, and other components of social-environmental systems are collocated and interact across ecological or social landscapes. The presence of wildlife corridors that link different patches of habitats are one such example of connectivity; communication networks and early warning systems that link people and communities are a second. In the conservation world there is considerable evidence that connectivity—for example, through corridors that link remnant patches of coral reefs or forests—is critical for recovery from disturbance and maintaining population viability and genetic diversity. Likewise, improved and increased methods of sharing information and new ideas have been shown to be essential to the development of solutions to social and technological challenges.

As with most things, it is possible to have too much diversity, redundancy, and connectivity, and end up with exactly the opposite result of the one expected. How much is too much, of course, depends on the particular system and requires contextual research and experience to disentangle. For our purposes, however, it is enough to realize that these structural properties matter and deserve attention in the pursuit of sustainability.

Given that social-environmental systems are so complex, and that vulnerabilities and tipping points abound, the knowledge capital that we discussed in chapter 2 is critically important. Because our best sources of reliable knowledge are not just science but also real-world experience in trying to manage for sustainability, the capacity to learn needs to be viewed as one of society's most important assets. **Learning** can usefully be thought of as the process of acquiring new or modifying existing knowledge, skills, and behaviors.[11] It is of fundamental importance in building resilient systems. Likewise, effectively managing a social-environmental system and maintaining or increasing its resilience to disturbance requires continuous revision and updating of knowledge. Experimentation and monitoring are examples of the way learning occurs; equally important are the ways that knowledge is produced and shared among researchers and decision makers. Chapter 5 will focus exclusively on this topic.

EVALUATING COMPLEX SYSTEMS

Given all these complications that play out in complex social-environmental systems, how can their dynamics be better understood and evaluated? Can the connections, interactions, and feedbacks be well enough mapped out to prevent unintended negative consequences or unpleasant surprises as a result of society's or nature's interventions? Unfortunately, the answer now and for the foreseeable future is no. Sustainability science is a young field and is far from having either a comprehensive, integrated understanding of social-environmental systems or the tools to evaluate them quantitatively. Nevertheless, we believe that deploying even a limited knowledge of such systems within the sustainability framework adopted for this book can change mental models and ultimately the ways in which decision makers manage for sustainability. Moreover, much more is being learned as the field continues to develop. Efforts that approximate and predict the dynamics and functioning of particular social-environmental systems in particular contexts and then follow the interventions with cycles of testing, learning, and revising are expanding our knowledge. It is therefore important to understand what kinds of models, tools, and approaches can help decision makers assess their choices and then monitor, evaluate, and learn from them.

A variety of such approaches allow analysis of social-environmental systems and the consequences and trade-offs that occur within them under different scenarios. While these approaches rarely encompass the entire social-environmental system, they can nonetheless be useful in articulating at least part of the system in the context of the whole. It is not the purpose of this chapter to provide in-depth background on these tools and approaches, but we hope the reader will be intrigued enough to pursue them in further study.[12]

Analytical Models

Within many fields, system dynamics modeling is used to conceptualize and then formalize an understanding of the interactions in a system. Analysts might start with a simple story or drawing that provides a qualitative picture of the system, its most important elements or components, and flows among them. Once they have sketched such a picture, the analysts can test their understanding of the system and explore hypotheses about how the system functions and how components of the system change in response to changing decisions or interventions.

Mathematical models that attempt to integrate social and environmental dynamics are increasingly becoming part of this analysis. For example, agent-based models mathematically represent the aggregate of decisions made by a group of individual "agents" as they use resources or allocate assets in a social-environmental system. Combined with spatially explicit information on the natural capital of the place (e.g., the land or resource base), as well as other assets, such models can be used to better understand the workings of the system and how it is likely to change over time as a consequence of management decisions or policies.

One modeling approach that is useful at regional to global scales is integrated assessment, used, for example, to project climate change and its impacts under various assumptions about the future. Using different scenarios of energy use, population dynamics, technology, and development trajectories, such models project future greenhouse gas emissions and use the resulting atmospheric concentrations of greenhouse gases to drive climate models, which are ultimately linked to social and environmental impact models. These linked models allow exploration of the effect of energy policy scenarios in terms of human well-being and of the assets of the productive base on which well-being relies. Importantly, these models are moving toward increasing integration of model subcomponents and improved ability to evaluate the coupled system.

A variety of ecosystem services models are in use or under development to help decision makers understand how management of natural capital assets affects the goods and services that can be produced using those assets. These models range from sophisticated simulation approaches to structured survey approaches. They are all intended to evaluate the effects of different choices on the value of ecosystem services produced. They are decision-support tools in that they provide decision makers with information about trade-offs and co-benefits that occur under different scenarios (e.g., different choices about how to use and manage land or water in a given place). Results from such analyses do not tell decision makers what to do but do provide them with information they may use to make choices in the context of their broader values and concerns (see box 3.1 for an example of an ecosystem services model that has made a difference for decision making).

Likewise, multi-criteria analysis (MCA) develops spatial data from satellite or ground-based databases on different land characteristics and overlays those data layers so that decision makers can identify options for the land uses that allow for the most benefits, or the lowest costs, in the context of their goals and values. For example, companies or municipalities selecting sites for deployment of solar arrays or wind turbines might use MCA to identify locations that are already limited

BOX 3.1. Evaluating Ecosystem Services for Decision Making

Hawaii's largest private landowner, Kamehameha Schools, invited the Natural Capital Project in 2007 to map and value ecosystem services on its unused agricultural land to help make decisions about the land's future. The Natural Capital Project (www.natural capitalproject.org), a partnership between Stanford University, the University of Minnesota, The Nature Conservancy, and the World Wildlife Fund (WWF), engaged with Kamehameha Schools and the surrounding community for two years. Working closely with the key stakeholders, scientists in the Natural Capital Project grew to understand the social-environmental system within which the lands were situated.[13] The project team was then able to create (using their modeling software known as InVEST) spatial models of what would happen to carbon storage, water quality, income earned, biodiversity, and other services under three different development scenarios (sugarcane monocrops, diversified agriculture and forestry, and residential development). These models were then shared with Kamehameha Schools and the affected communities so that they could discuss the potential trade-offs.

The models showed that the highest near-term income from the land would result from the residential development scenario, but

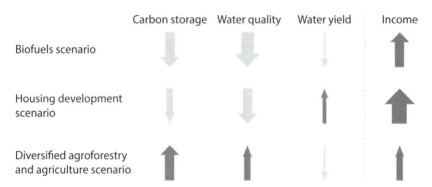

FIGURE 3.3. Projected changes in ecosystem services under three scenarios of future land use using the InVEST ecosystem services model on Oahu, Hawaii. Size of the arrow indicates relative size of the cost or benefit. (Modified from Goldstein, J. H., G. Caldarone, T. K. Duarte, D. Ennaanay, N. Hannahs, G. Mendoza, S. Polasky, S. Wolny, and G. C. Daily. 2012. "Integrating Ecosystem-Service Trad-eoffs into Land-Use Decisions." *Proceedings of the National Academy of Sciences of the United States* 109[19]: 7565–70; a part of the Natural Capital Project: www.naturalcapitalproject.org)

BOX 3.1. (*continued*)

that carbon storage and water quality would suffer. The sugarcane scenario actually showed the worst carbon storage rating (because of forest clearing that would have to occur) and also had water-quality impacts. The diversified agriculture and forestry scenario positively affected carbon storage and water quality but offered the lowest immediate income. In the end, Kamehameha Schools chose the diversified agriculture and forestry option, because it aligned best with their mission to balance environmental, economic, cultural, and community values. Besides the decision itself, the real benefit of the project was to bring out the key trade-offs "into the open," where the landowner and stakeholders could see and talk about them. In part because of the Natural Capital Project's work, Kamehameha Schools won the 2011 National Planning Excellence Award for Innovation in Sustaining Places from the American Planning Association.

---

with respect to biodiversity or conservation or cultural value but have the greatest resource (sun or wind) value, appropriate land ownership, and good access to electricity grids.

Long-term spatially explicit data on all components of the social-environmental system are crucial to constructing conceptual and mathematical models, empirically testing relationships, and tracking change as it occurs. Data acquired from ground-based, airborne, and satellite-based remote-sensing techniques have dramatically improved the measurement and understanding of regional- and global-scale changes in the biophysical parts of the system—including shifts in land use, soil and water quality and quantity, infrastructural patterns, ecosystem status, and many other variables. Data from ground-based observations and experiments can be related to these remotely sensed data and statistically evaluated, including in a spatially explicit context (for example, using geographic information systems [GIS]).

Likewise, data from corporate, local, state, and national surveys and assessments contribute to an understanding of social-economic parts of the systems. Census data and household surveys, along with aggregated data from a variety of other sources, provide information on health, education, population size and demographics, migration patterns, income, land tenure and ownership patterns, and many other variables.

Such data are less easily expressed spatially, and challenges in linking biophysical and social data often center on the mismatch in spatial and temporal characteristics of the databases. The development and linking of such databases, and the use of models and new "big data" analytics to interpret them, will ultimately improve the ability to evaluate and track trends in sustainable development.

Life cycle assessment (LCA) attempts to quantify the system impacts on the capital asset stock of alternative production processes. Its special contribution is in systematically mapping the implications of such processes "from cradle to grave." That is, LCA seeks to tally (1) how much the production process draws down the raw materials and other capital assets that it uses for inputs; (2) how much the process generates both goods and "bads" like pollution and wastes that damage the capital asset base; and (3) and the extent to which the goods so produced can be recovered at the end of their useful life to once again enhance the capital asset base.[14]

An LCA starts with setting out the goals and scope for a particular analysis and then generally includes three core stages: an inventory, in which basic measures of energy, raw materials, and emissions are calculated; an impact assessment, in which the inventory information is translated into the possible environmental or social harms they represent (e.g., logging forests for building materials can lead to habitat reduction and erosion); and an interpretation and analysis of results with recommendations for improvements in the manufacturing process or changes in public policy.[15]

LCA enables people to make more informed choices about the trade-offs among different products or processes, though it does not necessarily simplify those decisions. For instance, plastic bags last much longer in the environment than paper and can harm marine life, but they take much less energy and water to produce. Deciding which impacts are of greatest concern may require public discussions and values assessments, which in turn can be enhanced by appropriate investments in the relevant social capital.

Accounting and Indicator Systems

How can communities, corporations, governments, and non-governmental organizations (NGOs) predict and track their success in managing social well-being? In the last several decades, substantial effort has focused on the development of accounting systems and appropriate and reliable indicator systems for components of a system. Such accounting and indicator systems can estimate environmental damages, the value of natural resources, or national wealth and well-being using measures of gross domestic product (GDP) or gross national income (GNI). These and related measures tend to have at least two weaknesses from a sustainability point of view. First, they measure flows (what is happening now) rather than the stocks or assets that are central to sustainability (what's left to draw on in the future), analogous to flying across the ocean and monitoring only airspeed while neglecting the readout of the fuel gauge. Second, they fail to recognize and integrate the social and environmental as well as the economic determinants of well-being.

Measures of GDP and GNI, as widely used, suffer from other limitations as sustainability metrics. In general, they do not incorporate recognition of or account for externalities, nonmarket assets, wealth distribution issues, and future costs and benefits. They may fail, for example, to account for transboundary effects, as when one country's air pollutants are carried across national boundaries and harm another, or upstream resource use by farming communities affects downstream fisher communities. Importantly, these measures also exclude negative impacts of current activities on the well-being of future generations, such as degradation of natural capital and other capital assets. These indices represent a measure of the current situation, not of the sustainability of that situation.

Increasingly, however, accounting and indicator systems are becoming more integrative, and although the field is far from mature, progress is being made. In a thorough review of a subset of indicators that represented the breadth of diversity in the group, Parris and Kates evaluated the motivations and goals along with metrics and scale of operation of a large set of indicators.[16] Their review and others like it conclude that while indicators abound and are used widely by governments, corporations, and others—from local to global scales—there is little consistency among them: they are used for very different purposes, with different terminologies, data, and methods of measurement. Moreover, there is not much evidence that they have been used effectively over time to track progress toward sustainability goals. Nevertheless, decision makers are seeking ways of understanding the state of play for all the capital assets in our sustainability framework.

In recent years, a number of new types of social/economic progress metrics have been developed that show promise for measuring trajectories of inclusive well-being. For example, a team of scholars at Yale[17] developed a framework for integrating externalities (in this case, industrial pollutant emissions and the damages they cause to natural, human, and manufactured capital) into national economic accounts. They developed a "gross external damages" (GED) metric that shows how much harm is done by emissions from various industries and compared this measure with the market value (referred to as value added, VA) of those industries. The analysis (which we discuss in more detail in box 3.2, with data in table 3.1) shows that for industries with a GED/VA greater than one (such as solid-waste incineration and coal and petroleum-fired power plants), society is paying too low a price (assigns too low a value) for the goods and services produced by the industry. For rational policy, either the damages need to come down (e.g., by incorporating different fuels or technologies in the process of producing electricity) or the market price needs to go up (e.g., via taxes to reflect the social damages and lower consumption) or both.

BOX 3.2. Framework for Integrating Externalities

In their framework for integrating externalities into national economic accounts, Yale researchers developed a "gross external damages" (GED) metric showing how much harm is done by emissions from various industries, compared with the market value (VA) of those industries, that is, how much people are currently paying as consumers for what the industry produces. The analysis is backed by exquisitely detailed science that identifies the emissions of each industry at each place in the country, then distributes and transforms those emissions over the country using state-of-the-art atmospheric chemistry and transport models, and finally assesses damages by comparing pollutant deposition at each point around the country to "dose-response" curves showing the relationship between concentration and impact. They then sum these numbers across the entire country to compute the GED resulting from all the industry's air pollution. Finally, they create a GED/VA ratio (table

BOX 3.2. (*continued*)

3.1). (The damages to society from greenhouse gas emissions and the resulting climate change are not included in the numbers reproduced in table 3.1, though they are discussed in the original paper. Other pollution damages, e.g., those associated with water pollution by the same industries, are also not included. The GED/VA ratios reported in the table are therefore certainly underestimates of the true values.)

So what? For industries with a GED/VA of more than one, the straightforward conclusion is, "since you are in a hole, stop digging." That is, with each additional unit of production from those industries (e.g., electricity from coal), society emerges worse off than

TABLE 3.1. U.S. INDUSTRIAL PERFORMANCE: GROSS EXTERNAL DAMAGES (GED) INFLICTED ON SOCIETY BY AIR POLLUTION (IN $BILLION, 2000 PRICES) AND ITS RATIO TO VALUE ADDED (GED/VA)

Industry	GED/VA	GED
Solid-waste combustion and incineration	6.72	4.9
Petroleum-fired electric power generation	5.13	1.8
Sewage treatment facilities	4.69	2.1
Coal-fired electric power generation	2.20	53.4
Dimension stone mining and quarrying	1.89	0.5
Marinas	1.51	2.2
Other petroleum and coal product manufacturing	1.35	0.7
Steam and air conditioning supply	1.02	0.3
Water transportation	1.00	7.7
Sugarcane mills	0.70	0.3
Carbon black manufacturing	0.70	0.4
Livestock production	0.56	14.8
Highway, street, and bridge construction	0.37	13.0
Crop production	0.34	15.3
Food service contractors	0.34	4.2
Petroleum refineries	0. 18	4.9
Truck transportation	0. 10	9.2

Source: Muller, N. Z., R. Mendelsohn, and W. Nordhaus. 2011. "Environmental Accounting for Pollution in the United States Economy." *American Economic Review* 101(5): 1649–75.

if the production hadn't happened at all. Industries in this category are the ones at the top of the list in the GED/VA rankings of table 3.1, most notably, waste treatment and electric power production based on coal or petroleum fuels. (The GED column shows that by far the biggest factor is excessive production of electricity from coal.) Sane people (or societies) would stop doing—or do less of— things that hurt them. More precisely, this rigorous research shows that for industries with a GED/VA value greater than one, given the damages caused by each additional unit of production from that industry, society is paying too low a price (and assigns too low a value) for the goods and services it produces. For rational policy, either the damages need to come down (e.g., by incorporating different fuels or technologies in the process of producing electricity), or the market price needs to go up (e.g., via taxes to reflect the social damages) or both. In either case, the (much) higher market price (from required technologies, or an emissions tax) should lead to (much) lower emissions and GEDs. At some point, the price would be sufficiently high, the technology sufficiently clean, and the demand sufficiently low that society would not perversely destroy its own well-being to continue utilizing the industry. To get to that point, however, will require that (1) policy making recognizes that "free markets" often create the sort of externalities that the Yale team so carefully documented; (2) scientific research to make those externalities "visible" to producers and consumers is essential; and (3) turning visibility into real corrective action requires governance structures that protect the rest of society from those who deplete inclusive well-being.

At the scale of national accounting, the Inclusive Wealth Project is of particular interest and promise. The project is an international UN-based effort to extend an approach initially developed at the World Bank to measure changes in the capital asset base of countries through time.[18] In particular, it develops an Inclusive Wealth Index (IWI) that is intended to show how an aggregate measure of the social value of society's capital assets changes over time. The theory behind the IWI— like that behind the sustainability framework presented in this book— holds that so long as the per capita inclusive wealth of society is not decreasing, neither is its per capita inclusive well-being, and thus its development can be tentatively judged to be "sustainable." The project's *Inclusive Wealth Report 2014: Measuring Progress toward Sustain-*

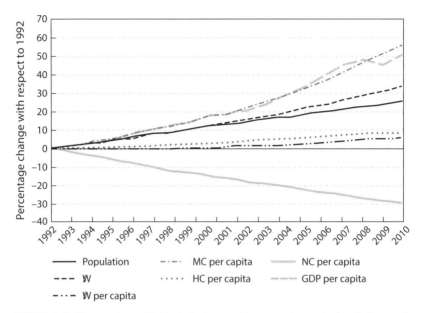

FIGURE 3.4. Changes in worldwide inclusive wealth per capita and other indicators for 1992–2010. Lines represent different types of capital: HC, human capital; MC, manufactured capital; NC, natural capital; and GDP, gross domestic product. *W* is the inclusive wealth index. (UNU-IHDP and UNEP. 2014. *Inclusive Wealth Report 2014: Measuring Progress toward Sustainability*. Cambridge, MA: Cambridge University Press)

ability[19] compiles data on a selection of the natural, human, and manufactured capital assets of 140 countries over the period 1990–2010. (Although the current approach is far from complete, ongoing work of the project seeks to incorporate additional dimensions of the assets already included and to address changes in social and knowledge capital.) Some of the results published in the 2014 report are reproduced in figure 3.4. These suggest that the development trajectory for the world overall is (barely) increasing its inclusive wealth per capita and thus is (barely) sustainable. Not surprisingly, the IWI data on trajectories of individual regions and countries suggest that some are on development paths that are comfortably sustainable (e.g., Western Europe), while others are on paths that—if continued—will leave their grandchildren impoverished (e.g., East Africa). Less obviously, it highlights for each country which capital assets are on trends that most support sustainable development, and which are on trends that undermine it. The IWI is, of course, incomplete. It explicitly tracks only the "tangible" capital assets (natural, human, and manufactured or, as IWR calls it "produced") while estimating the difficult-to-measure "intangible" assets of social and knowledge capital. And, constrained by limitations of data and theory, the IWI includes only a few of the many elements that should eventually be enumerated in calculating values of the broad tangible asset stocks. It therefore seriously underestimates the true wealth of nations, though the trends it reports are not obviously biased in one direction or another. Nonetheless, the IWI is a beginning, and for pursuing sustainability it is a huge improvement over previous metrics such as the GNP or the human development index (HDI). A vigorous program of research and monitoring is under way to improve future editions of the report.

THE NEED FOR INTERDISCIPLINARY
RESEARCH AND PROBLEM SOLVING

Reflecting on what we've discussed in this chapter, it is easy to see why understanding and managing coupled social-environmental systems for inclusive well-being is such an enormous challenge. These are complex systems with many interactions among many parts, and analytical frameworks for comprehending them are in early stages of development. To understand social-environmental systems, and to pursue sustainability within them, researchers, decision makers, and concerned citizens have to draw on many kinds of knowledge and know-how. Science for sustainability is a big tent, and many different kinds of expertise are needed (recall the range of scholarly disciplines and professions listed earlier in table 2.1).

However, all these different kinds of expertise and knowledge would fall far short if they operated alone. Researchers and practicing professionals need to understand each other's languages, respect and appreciate each other's methods and approaches, and recognize where and how their limited knowledge can make a difference. Ultimately, understanding and managing social-environmental systems for sustainability requires team efforts with both disciplinary expertise and specialists who work purposefully at the interfaces among components of these complex adaptive systems. We illustrated the challenges and rewards of such teamwork in our case study of the Yaqui Valley. We dig more deeply in chapter 5 into what is known about the kinds of partnerships among decision makers and researchers that can successfully link knowledge with action for sustainable development. First, however, we turn in chapter 4 to the governance systems within which those partnerships must operate and through which experts—and also the other "actors and agents of change" highlighted in our framework (figure 2.1)—must learn to work in their pursuit of sustainability.

CHAPTER 4

Governance in Social-Environmental Systems

The concept of **governance** includes the decision-making process that creates and enforces rules about what people may, must, and must not do in interactions with one another and with the rest of the social-environmental system.[1] In the context of the framework for understanding sustainable development and working toward sustainability that we use in this book (figure 2.1), the governance process, and the rules that it produces, are an important part of a society's assets of social capital. Governance and the resulting rules might address very concrete questions, such as, How much pollution is a company allowed to emit? Who is entitled to unemployment benefits? What is the punishment for breaking a given rule? The rules will also define who has the right to participate in governance decision making and who is in charge of enforcing the rules.

When the social "actors and agents of change" of our framework agree to create, implement, and abide by rules, they are thus engaged in processes of governance. Those **governance processes**, together with other production and consumption processes we discussed in previous chapters, determine the dynamics of social-environmental systems.

Governance is important in the pursuit of sustainability in that its rules or arrangements are among the core components of social capital and thus are among the determinants of inclusive wealth and well-being. Governance processes provide a means through which people can change the nature of social-environmental interactions so that human activities do not drive down the overall asset base on which future generations depend. Through a mix of "carrots and sticks," governance can produce rule systems that sway citizens, corporations, governments, and other organizations to contribute to the goals of sustainable development.

To govern a social-environmental system effectively, however, is easier said than done. Pursuing sustainability will require better integration, more coordination, and cooperation at an unprecedented scale. Many of the problems discussed in this book—such as overfishing, rural poverty, and pollution—have come about because of limited or failed efforts to govern our shared social-environmental systems. Societies with weak governance processes are prone to generate production and consumption patterns that degrade capital assets while improving the well-being of only a select few members of those societies.

It is not just the *lack* of governance that poses a problem for sustainable development. Sometimes governance itself is the problem. One of the challenges in creating governance arrangements more consistent with the goals of sustainable development is that the actors most influential in shaping the existing processes—such as nation-states, corporations, political parties, and sometimes special-interest groups—rarely hold the same goals and often disagree on how social-environmental systems ought to be governed. Some of these actors benefit enormously from unsustainable practices and resist any effort to change business as usual because such changes might jeopardize their privileged positions in society. Individuals and organizations pursuing sustainability thus face a serious challenge: how can the existing governance processes and arrangements, which are shaped by powerful actors with vested interests, be reshaped so that they are more conducive to achieving inclusive social well-being?

Strategies for overcoming such challenges can benefit from a deeper understanding of governance and how it can promote social change. Simply saying "we need to change things" or "we need to cooperate more" is rarely helpful. More nuanced strategies that address some of the fundamental barriers to social change are needed. The goal of this chapter is to outline some of the concepts and lessons learned about understanding and influencing governance in social-environmental systems. It ends with a discussion about how a better understanding of governance can help the quest to organize concerned citizens and scientists in ways that contribute to the pursuit of sustainability.

CONCEPTS AND A FRAMEWORK FOR ANALYZING GOVERNANCE

Several conceptual foundations underlie our understanding of what governance is and why it is important for sustainability. In this section, we discuss these concepts and propose a conceptual framework for diagnosing governance problems and for exploring possible options for addressing such problems.

Core Governance Concepts

Three concepts are useful for developing a better understanding of governance and how it affects sustainable development efforts: (1) collective action and collective-action problems, (2) externalities, and (3) common-pool resources and the "tragedy of the commons."

Collective action occurs when two or more individuals cooperate to accomplish a goal they cannot achieve individually. For example, the group of farmers described in our Nepal case engaged in collective action when they decided to work together to build and maintain their own irrigation system. Groups of citizens concerned about loss of ozone acted collectively to urge their government leaders to keep the negotiations leading up to the Montreal Protocol moving forward. A group of students act collectively when they organize a campaign to put pressure on their university to adopt a zero–carbon emission goal. Many of the traits or "goods" of a functioning society—such as law and order, public safety, economic growth, and environmental protection—require coordinated collective actions at a very large scale. Because these "goods" will not evolve automatically, it is necessary to develop a structured process for agreeing on and implementing rules to steer individual behavior in ways that promote outcomes that are good for society as a whole. We refer to this decision-making process as governance. Governance is an important part of a society's social capital because it can produce the rules needed to address collective-action problems.

A **collective-action problem** occurs when a group of individuals fail to accomplish their common goal. People can fail to cooperate with one another for many reasons. They may lack motivation to cooperate or lack the information needed to accomplish the goal, or they may engage in a power struggle over who should control the decision-making process. Collective-action problems are common in all efforts to make changes in society. For example, students who are mobilizing to persuade their university's leadership to act on climate change will try to get other students to participate in the campaign, but some of those other students may not be motivated to join the collective action. In that particular case, there are several possible causes for the collective-action problem: students may not believe that the protest will actually make a difference, or they may have other activities that are more important to them, or they may think climate change is not caused by humans, or they may agree with the greater cause of the protest but prefer that others do the protesting. Student organizers need to figure out a way to overcome these collective-action problems, or their efforts will fail.

Sometimes, the failure to act collectively leads to individual actions that have significant consequences for people who are not directly involved with the decisions. When these consequences cause damage, they are referred to as **negative externalities**, because decisions adversely affect third parties who are external to the setting in which the decisions were made. As we discussed in chapter 3, this is one form of the "invisibilities" common to complex social-environmental systems. Pollution is a classic example of a negative externality. Because the polluter—a business, for example—does not bear the full cost of its actions (in an unregulated scenario), it is therefore likely to pollute more than would be the case if it had to take the costs of the damages into account.

Negative externalities are particularly difficult to address when the affected party is not well organized politically or has limited opportunities to influence the governance process that could potentially correct the externality. Nevertheless, substantial progress has been made in the governance of the use of some resources susceptible to negative externalities. Harkening back to our case studies, the damage to the stratospheric ozone layer caused by the production and consumption of CFCs has been effectively addressed by a global treaty. Forceful governance responses to the Great Stink of London managed to clean the Thames

River of many of the wastes it collected as it passed through the city. These examples are a historical record of real progress on governance in social-environmental systems, although there is much more to be done.

Positive externalities also exist. The form most relevant to sustainable development may be technical innovation. Inventors often see the fruits of their work adopted by others who did not bear the costs and risks of research and experimentation. While such widespread, relatively costless adoption can be an effective way of assuring that the benefits of a particular innovation are accessible to all (especially the poor), it can also undercut the profit motivation for future inventions. Managing such positive externalities—assuring that inventors will continue to have an economic incentive to produce the innovations societies need but also that the results of their labors will be widely accessible on terms that everyone can afford—likewise requires creative governance processes to promote collective action. An early response to such challenges was the system of international agricultural research centers that invented and then gave away for free the seeds that launched the Green Revolution. The centers could do this in part because they were staffed by idealistic researchers and funded by public-spirited private foundations and, later, by national governments working through the UN system and the World Bank.

In a more recent situation, effective drug therapies for HIV/AIDS were finally created through investments of rich governments and profit-making pharmaceutical firms. But firms seeking to recoup their research costs filed patents on the drugs and then charged prices that could not be afforded in poor countries—a classic positive-externality dilemma. The tension between the need to incentivize future research and the responsibility to provide access to essential drugs presented a dilemma for governance. That dilemma was resolved, however, after political clashes between international civil society networks and developing-country governments on one side and the pharmaceutical industry on the other gradually produced an unwritten political bargain: wealthy countries would continue to pay high prices for HIV medicines to reward research and development investments, while developing countries would purchase generic drugs at close to the cost of production through exceptions made to patent rules.[2] More such inventive governance arrangements—which almost always involve collective action among activists, entrepreneurs, scientists, and lawyers—are needed in the pursuit of sustainability.

One of the most well-known collective-action problems related to social-environmental systems is described in Garret Hardin's 1968 classic paper, *The Tragedy of the Commons*. Hardin argued that people who share resources are trapped in a collective-action problem that produces competition for resources that can spin out of control until the resource eventually collapses.[3] The analogy he used in his paper is that of two farmers who share a meadow where they graze their individually owned cattle. Each farmer sees it in his own interest to keep adding more animals to the shared meadow. And even though the farmers know they would both be better off in the long run if they maintained a limited and stable population of grazing animals, they are unable to reach such an agreement, leading to a degradation of the grazing system and a bad outcome for both.

What Hardin describes is a specific type of collective-action problem that sometimes occurs in situations where multiple individuals share a **common-pool resource (CPR)**. Many of our natural resources such as clean air, forests, and groundwater may be considered CPRs. They are particularly challenging to manage, because they are finite resources, and it is difficult to prevent anyone from using them. Because of these attributes of CPRs, a rule system that is recognized by the resource users is needed to control access to and regulate use of the shared resources. In that sense, governance can be thought of as a strategy for addressing behavioral problems that threaten the condition of the shared resources. If such arrangements are not created and enforced, the CPR is at risk of overuse, and a tragedy of the commons becomes more likely.

In a nutshell, the significance of governance in the social-environmental system is that it can help overcome collective-action problems in society by producing rule systems that steer individual actors (people, corporations, states) so that their behavior is consistent with the long-term goals for a society as a whole.

Unpacking Governance: A Conceptual Framework

Conceptual frameworks and models help analysts think more systematically about complex issues. In chapter 2 we discussed a broad framework for thinking about sustainability goals and linking them through the dynamics of the social-environmental system to the capital assets that are their underlying determinants. That framework identified governance as an important component of the social-capital assets. While the framework in chapter 2 is comprehensive, a more detailed framework "nested" within it is helpful for assessing the specific factors and characteristics that influence the ability of governance to alter existing unsustainable behaviors. We present such a framework in figure 4.1 and hope it will help the reader appreciate how governance matters for sustainable development, as well as diagnose potential shortcomings related to governance in specific contexts.

Our governance framework consists of six main components. Starting from the left-hand side in figure 4.1, the first component is *actors and agency* (compare with the leftmost box of our more general sustainability framework in figure 2.1). These are the people, groups, organizations, or states involved in the decision-making process or affected by the decisions. **Actors** can be as large as countries or multinational organizations and as small as households or individual people. Their particular characteristics include values, beliefs, power, agenda, interests, capacity, and motivation. One of the goals of this part of the analysis is to assess the degree to which different actors have **agency**, that is, the degree to which individual actors have the capacity to act independently and to make their own decisions. Agency is an abstract concept that is difficult to measure directly, but it is possible to characterize an actor's degree of agency relative to that of other actors by comparing access to political power, financial resources, and information. In sum, this stage of the governance analysis seeks to identify the main actors, their interests, and main characteristics—and how these affect their degree of agency.

In the Yaqui Valley case, researchers eventually mapped the actors who were associated in some way with decision making related to farming practices (see appendix A, figure A.8). In doing so, they discovered that it was the lending policies of one category of the non-individual-farmer actors in the valley—the farm credit unions—that effectively discouraged farmers from adopting more sustainable fertilizer techniques. Once this actor was identified as being an important influence on local farming practices in the valley, targeted efforts were made to understand the credit unions' concerns and to engage them in the creation and testing of technologies that could lead them to policy change.

Mapping out the actors, how they interact with one another, and assessing the main attributes of the political process can be helpful for understanding the sources of existing problems of sustainability as well as for identifying potential levers for change. Another benefit of mapping out the actors—the who's who in a particular context—is that it can help locate potential allies with whom to collaborate to become more effective in the decision process that may eventually produce change.

In the next step of the governance framework, the focus is on the ways in which the identified actors interact with one another through a **political process** in which collective decisions are made. As shown in figure 4.1, this political process can exhibit varying degrees of transparency, accountability, representation, and legitimacy. The knowledge gained from this stage of the governance analysis is often important for assessing the degree to which these existing processes are appropriate for addressing a particular problem. For example, in the case of the Nepali irrigation systems, the national government of Nepal decided in the 1960s to invest in the expansion and improvement of the country's irrigation infrastructure. However, the decision-making process for designing many of these systems involved mostly engineers, government officials, and international donor representatives. The political process failed, at least initially, to involve and be accountable to the farmers themselves, and this shortcoming led to poor design decisions (see appendix A for more details).

It is important to recognize that political processes exist at multiple levels and that decisions at one level are often influenced by decisions and governance processes at higher levels of political authority. For example, the rules created by the governance process at the national level may be influenced by international treaty negotiations, as was the case for the ozone treaty. Governance (and for that matter, most other social-environmental system processes) is often a nested enterprise.

One of the essential outcomes of the political process is **institutional arrangements**, the next stage in the governance framework. Institutional arrangements, or rules, can be either formal or informal.

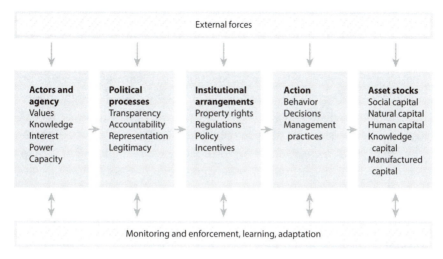

FIGURE 4.1. The governance framework for social-environmental systems. This is a higher-resolution framework to complement that introduced in figure 2.1. Its purpose is to help in the assessment of the effect of governance on human interactions with capital assets.

Examples of rules include policy, regulations, local norms and customs, contracts, and property-rights arrangements. Rules specify not only the rights and responsibilities for using and managing a resource but also the agents whose responsibility it is to monitor and enforce rule compliance.

This may be the most important part of the governance analysis, because so much of the effectiveness of the human responses to a problem depends on the specific characteristics of the rules introduced and how they are enforced (if at all). The farmers in Hardin's *The Tragedy of the Commons*, for example, were not constrained by any regulations or policy about how many animals they could add to the meadow, and no effective property rights to the land or the resources appear to have been enforced in that case. As a result, the farmers pursued their narrow, short-term self-interest until their social-environmental system collapsed.

One intention of introducing new institutional arrangements is to affect people's **incentives**. In governance analysis, the term *incentive* refers to expected rewards and punishments that individuals perceive to be related to their actions and those of others.[4] Rules define when, how, and to whom different rewards or punishments should be applied. Rewards can be monetary, such as a wage, tax credit, rebate, or a bonus. Rewards can also be nonmonetary in nature and include earning the respect of one's peers, the satisfaction from learning new skills or knowledge, affection, and the feeling of having done the right thing. The threat of punishments of different kinds can also function as incentives. Such punishments may be the threat of a lawsuit, consumer boycott, fine, jail sentence, social exclusion, or loss of employment. These are all external stimuli that may encourage more of some kinds of behavior and less of others.

In the subsequent stage of the governance process, the focus is on human **action**, including the behavior, decisions, and day-to-day practices that people choose to pursue. For institutional arrangements to affect these decisions and actions, people need to know about the arrangements, understand them, and choose to abide by them. That is unlikely to happen unless the governance processes enable actors to monitor and enforce rule compliance. For example, in the Nepali case, the local rule system laid the ground rules for what each family was expected to contribute for the building and maintenance of the irrigation canals and what would happen if an individual did not meet those expectations. The village elders monitored and enforced the agreements. It is hard to imagine that the villagers would have been able to build the canals, maintain them, and pay off the loan on time had it not been for the enforcement of these elaborate rules.

In the analysis of governance in social-environmental systems, special attention is paid to institutional arrangements that seek to regulate behavior related to human use of the asset base. As a result, it is useful to consider how the different **capital assets** are likely to respond to the introduction or modification of a particular institutional arrangement, which in turn will sway an actor to decide to take a certain action or behave in particular ways. These actions may have a direct effect on asset indicators, such as pollution levels (natural capital), people's health (human capital), the condition of road infrastructure (manufactured capital), the level of interpersonal trust in a society (social capital), and the creation of new inventions (knowledge capital).

The different uses and exploitations of the capital assets permitted or forbidden by institutional arrangements can lead to very different outcomes in the condition of these assets. Simply put, *the condition of the capital assets is unlikely to improve unless a critical mass of actors work together to change the rules about how the asset base is accessed and used.*

External forces also shape governance. At all levels of governance in a society, there are factors beyond the control of the actors at any given level (e.g., household, community, region). Examples of external forces include solar radiation, climate change, global population growth, natural disasters, global price fluctuations, decisions higher up in the political hierarchy, international agreements, and any other factor external to the governance system at any given level. Some of these forces are surprises, while others are predictable trends. When there is no apparent way to change these forces through the governance process at their scale, actors' best response to these external forces may be just to accept that they exist and adapt to them as best they can. At other times, it may be possible for actors to appeal to higher levels of governance, where collective action among actors may be able to respond more effectively to these external forces.

Each of our four cases described different types of external forces that affected how people interacted with the rest of the social-environmental system. For example, in the Nepali irrigation case, changes in climate—more frequent droughts and irregular monsoons—represented very disruptive external forces that prompted the local farmers to take collective actions. In the London case, the various epidemics that wreaked havoc among the city's dwellers were caused by a combination of the arrival of disease organisms from the outside (e.g., the plague and cholera) and the existing socioeconomic and biophysical conditions in London that facilitated the spread of the diseases.

But external forces can also have a positive effect, as was illustrated by the creation of chemical refrigerants to replace the ozone-depleting CFCs. This innovation—which perhaps wasn't completely external to the governance system, because the innovator (e.g., DuPont) was arguably one of the key actors—was nevertheless very important for the outcomes because it allowed people and companies to satisfy their demand for inexpensive coolants and aerosols without inflicting more harm on the stratospheric ozone layer.

The box at the bottom of figure 4.1 depicts feedback loops to monitor and enforce the governance process, to learn from experience, and to adapt future iterations of the governance process. In chapter 3 we discussed feedback loops as critical aspects of complexity in social-environmental systems. They function similarly here in reference to the governance process. In cases where actors are surprised by the way the governance process plays out—perhaps the new institutional arrangements didn't affect human behavior as expected—these actors may want to reassess the appropriateness of those arrangements. This adaptation and learning process may continue over time with varying degrees of success, as was clear from all the case studies introduced in chapter 1. As one hurdle is cleared, another often appears, and so the work to promote sustainable development continues.

The real challenge in applying this conceptual framework for governance is to determine which particular approaches to governance will work in specific contexts and situations. This framework will not produce one right answer on how to improve governance. It can, however, help analysts systematically assess the strengths and weaknesses of a given governance process and formulate more thoughtful strategies that can serve as the starting points for governance reform. Practitioners trying to develop such strategies will also benefit from several take-home lessons gleaned from contemporary research on governance.

LESSONS FOR GOVERNANCE IN SOCIAL-ENVIRONMENTAL SYSTEMS

Research on governance in social-environmental systems has produced several potentially useful findings applicable to governance reform efforts. The following five of those lessons are discussed next: (1) diagnose barriers to cooperation; (2) figure out what makes people tick; (3) assess multiple possible governance responses; (4) consider interventions to be experiments to learn from; and (5) recognize that knowledge is power.

Diagnose Collective-Action Problems

Before trying to fix something, one needs to understand what is broken. Understanding what the problem is and why good outcomes are not materializing is the most important goal of any governance analysis. A collective of any size—whether a local user group or a nation or even the global community—that is falling short of its expectations to achieve certain collective outcomes needs to understand why collective action is failing before it can design an effective solution. All efforts to mobilize collective action face a large variety of challenges, which may be grouped into motivational problems, asymmetries of power, and information problems. A brief description of each of these common barriers to collective action follows.

One of the most common sources of breakdowns in collective action is related to weak individual motivation to contribute to the common good. The better off an individual is in the current system, the weaker his or her motivation to invest in a process that might change it. Even if an individual would like to see a change in the status quo, that person is tempted to let others do the heavy lifting in the change process. **Free-riding**, or attempting to benefit from a shared good or service without contributing to its provision, is a common barrier to cooperation. The larger and more heterogeneous the group, the harder it is to overcome this barrier.

But people may not be motivated to engage in cooperation for many other reasons. They may not see the proposed cooperation as relevant enough to warrant their support or may not perceive the real costs of continuing with business as usual. They may be oblivious to the risks associated with their existing behavior and fail to see the looming crisis. Most of these motivational barriers have to do with the beliefs and values people hold and the ways in which people perceive themselves and the world around them. In our Nepali case, farmers shared a perception of crisis and high costs of inaction that ultimately brought them together. In essence, they were motivated to set aside their differences and instead focus on their common interest in building a common irrigation system. Actors who seek to facilitate collective action will be more effective if they understand the sources of people's motivation, or the lack thereof.

Asymmetries of power exist when some actors are more powerful than others. The basis for holding power can be economic, social, or political. Most collective-action situations occur in contexts shaped by preexisting distributions of power, in which elites may make decisions that give them a disproportionate share of available assets, and they resist any initiative that seeks to change the status quo even if such an initiative would mean higher productivity for society as a whole.

What implications do such asymmetries have for sustainable development? When actors do not have a say, or even a seat, at the decision-making table, they lack power in the governance process. As a result, the needs of marginalized groups may not be taken into account when collective decisions are made. Regarding sustainability, future generations face similar asymmetries, because they have neither money nor political clout and thus can have a seat and a voice only through people today who take it upon themselves to speak on their behalf.

While asymmetries of power remain a significant barrier to governance reform and collective-action efforts in many societies, several recent developments demonstrate that they can be overcome. For example, in several Latin American countries, a deepening of the democratic process has allowed indigenous groups to reclaim ownership of significant parts of their ancestral lands.[5] A second example relates to persons who are displaced by the building of dams. The World Commission of Dams—a group created in 1997 consisting of members from civil society, academia, the private sector, professional associations, and one government representative—was charged with assessing the social, economic, and environmental impacts of large dam projects around the world and issuing a series of recommendations. Their first very influential report, published in 2000, made ten recommendations that are still considered the gold standard for how large dam projects ought to be conducted.[6] Among the commission's recommendations are those calling for protecting the rights of the people who would be displaced by the dam projects and allowing such people to have a say in the decision-making process.

Collective action may also break down because of informational problems. These arise in situations where actors do not share the same information about available choices, the consequences of those choices, or the characteristics and preferences of other actors with whom they interact. As we saw in the case of farmer-managed irrigation systems in Nepal, ill-designed government irrigation projects that failed to make good use of the farmers' local expertise led to decreases in productivity.

Once again, however, information problems can be overcome. The ozone case provides an illustration of how international scientific assessments can, if sensitively conducted, do the job. In the wake of the discovery of the Antarctic ozone hole in the mid-1980s, a controversy arose over whether significant declines in stratospheric ozone were occurring globally. Different models and data sets gave different results. The potential existed for an information war in which various stakeholders (countries, industries, environmentalists, etc.) would have been championing—or been expected to champion—the data set and models that most supported their own policy preferences. Instead, however, the U.S. National Aeronautics and Space Administration (NASA) established the International Ozone Trends Panel with a goal of making an independent and authoritative assessment available to all. The organizers included on the panel a diverse group of recognized experts drawn from ten different countries and from a cross section of industry, academia, and government. The panel members, who participated as scholars rather than representatives, managed to produce a rigorous consensus report showing that a significant downward global trend in stratospheric ozone was indeed under way, and that it was almost certainly caused by human emissions of ozone-depleting chemicals. The panel's report was highly influential, cited by government and industry leaders alike with providing a common foundation of scientific information that supported and necessitated action. The inclusive approach of the panel in producing this assessment that could then support governance processes has since been widely emulated, for reasons we explore in more depth in chapter 5.[7]

Understanding which particular collective-action problems are threatening human well-being provides the basis for devising more effective governance strategies. The more precise this diagnosis is, the more targeted the governance response can be. The chances of overcoming problems are greatly enhanced by an appreciation of the factors that motivate different governance actors to come together and actively contribute to governance responses that represent the common good. In other words, understanding what motivates individual group members allows groups to invent incentives to encourage engagement and cooperation.

Figure Out What Makes People Tick

Research on human behavior distinguishes between two sources of motivation: intrinsic and extrinsic. Intrinsic sources of motivation include moral and spiritual values, personal freedom, as well as a sense of autonomy. Extrinsic sources of motivation include cash payments, threats of punishment, and social exclusion. Most public policies make use of extrinsic sources of motivation to sway the behavior of individual members of society in a predetermined direction. If you get caught running a red light, you will have to pay a hefty fine and may even have your driver's license temporarily suspended. If you buy a fuel-efficient car, your government may give you a tax credit. These extrinsic sources of motivation often work quite well to induce desired group behavior, but recent scholarship has shown that conventional policy instruments that rely on top-down solutions and cash payments can sometimes be problematic. The following three studies illustrate this clearly.

Experimental evidence suggests that paying people money to encourage them to contribute to the common good can backfire and sometimes actually produce *less* willingness to contribute. Bruno Frey, a Swiss economist, conducted an experiment in which he asked a sample of Swiss citizens about their willingness to host a nuclear waste facility in their municipality under two different conditions.[8] He started by asking them the straightforward question whether they would be willing to host the facility in their municipality, and he found that a little over half of the respondents said yes. As a follow-up question he asked respondents whether they would be willing to host the facility if the central government paid them a substantial amount, between USD $2000 and $6000, as compensation for their services to the public. What is quite astonishing is that the percentage of citizens who were willing to host the facility dropped by half when a cash payment was offered. What explains this result? When asked, citizens' answers ranged from "hosting the facility is the right thing to do" to "the government cannot bribe us to do this" to "our support is not for sale." The finding suggests that use of cash payments can sometimes weaken people's motivation to do what is best for society.

Just as cash payments can backfire, cash fines can also be problematic. An example comes from a study of day-care centers in Haifa, Israel, that tried to make parents show up on time to pick up their children.[9] To do so, the day-care centers introduced a small fine of three dollars each day parents showed up late. The surprising result was that after the fine was introduced, *more* parents started to show up late, not fewer. Even more surprising, when the centers stopped issuing the fine after twelve weeks, the number of latecomers did not change at all. Before the fine was introduced there was only social punishment for late pickups: the day-care personnel would simply acknowledge the parents' tardiness and perhaps criticize them for making the personnel work longer hours. Once the parents paid the fine, however, it seems most of them no longer viewed themselves as ethically obliged to show up on time: they may have felt they had bought the right to show up late. In this particular situation, the social punishments produced more compliance than the relatively small cash fines.

Colombian economist Juan Camilo Cardenas conducted an experiment with members of rural communities in Colombia. The experiment sought to reproduce a recurring collective-action problem that these communities face in the management of forest resources.[10] In the experiment the villagers were asked to make individual harvesting decisions. The payoffs offered to participants depended on their performance in the experiment and were designed to capture the tension that exists in the real world between what is best for the individual in the short term and what is best for the group as a whole over the longer term. At the start of the experiment no communication was allowed, and participants ended up harvesting at unsustainable rates. The experimenters then changed the rules, introducing two different interventions to deal with the overharvesting problem, applied to two different groups participating in the experiment. One of the interventions allowed participants to communicate with one another before making their individual harvesting decisions, and the other simulated a government-imposed regulation: participants were not allowed to harvest above a certain level. Their behavior was monitored, and if they were caught overharvesting, they were issued a fine. The result: the participants that were permitted to communicate sacrificed five times more of their personal short-term profits to protect the forest than participants who were regulated by the government. In fact, the government intervention led participants to overharvest even more than during the first eight rounds, when there was no communication and no regulation.

These research findings highlight three things about the challenge of developing effective governance strategies to encourage people to pursue sustainability in their production and consumption decisions. First, the factors that motivate human behavior vary from one context to the next, but people often respond to a mix of both intrinsic and extrinsic sources of motivation. Second, the effectiveness of using extrinsic incentives—such as fines or cash awards—often depends on individuals' social norms, beliefs, and values. And finally, interventions are more likely to succeed when they acknowledge the autonomy and problem-solving capabilities of people whose behavior the interventions are trying to change.

Explore Multiple Governance Responses

In *The Tragedy of the Commons*, Hardin suggested that the only way to avoid the tragedy of overharvesting would be for a governmental authority to intervene. The outside intervention, he suggested, could happen in one of two ways—either by having a governmental authority coerce the resource users to limit their resource use or by privatizing the resource to individual owners, who would then see it in their individual self-interest to manage the resource more sustainably. For years, these two options dominated the policy instruments used to deal with common-pool resource problems. More recent scholarship has shown, however, that there is a third policy alternative that does not necessarily call for a top-down intervention à la Hardin: local resources users can, and often do, craft their own, home-grown solutions to commons problems. There is evidence that not only are locally crafted arrangements possible, but they sometimes even outperform national governments in natural resource governance.

Research on the governance of common-pool resources provides several lessons on what works and what doesn't when groups of people—from small local groups to the global community—try to manage fragile resource systems. A summary of three key findings follows.

Elinor Ostrom, who won the Nobel Prize in Economics in 2009 for her work on commons governance, examined hundreds of cases that involved local users seeking to manage a common-pool resource. Looking at both failures and successes, and systematically examining a host of factors that could help explain the variation in local outcomes, she found that there are a limited number of specific conditions under which resource users are more likely to be successful in coming up with their own institutional arrangements that can sustain their shared resources.[11] She proposed a list of eight principles, often referred to as "Ostrom's design principles":

1. The boundaries for the resource *and* user-group rights are clearly defined.
2. The rules governing use of commons match the local needs and conditions.
3. Those affected by the rules can participate in decisions about those rules.
4. There is an effective local monitoring system, in which local community members monitor one another's behavior.
5. There are graduated sanctions for people who don't comply with the rules.
6. Mechanisms for conflict resolution are cheap and easily accessed.
7. Higher-level authorities respect the autonomy and self-determination of the local user group.
8. In the case of larger common-pool resources, organization is in the form of multiple layers of nested enterprises, with local user groups at the base level.

Since Ostrom's original study was first published in 1990, hundreds of empirical studies have been carried out to test her original propositions in a variety of contexts. While these studies have produced some refinements to the original design principles, their core ideas have withstood the test of time.[12] Local self-governance is possible but is always a challenge.

A number of empirical studies from developing countries suggest that local governance processes can be more cost-effective than top-down governmental interventions in managing natural resource systems. These studies compare the performance of national governmental programs with local voluntary community-based initiatives to manage shared natural resources, such as forests and rivers. In one such study, Somanathan and colleagues collected data on the cost-effectiveness of different approaches to forest conservation activities in a large number of forests in India.[13] They found that local communities, on average, are better at conserving forests in terms of maintaining stable forest conditions compared with governmental organizations, and they do so at a fraction of the cost of the governmental programs. The homegrown, bottom-up approach to environmental problem solving is sometimes incomplete, however, especially when it comes to dealing with problems that are larger than the reach, or jurisdiction, of the local governance system (recall our invisibilities problem discussed in chapter 3). To address those broader problems, multilevel governance approaches may be needed.

Multilevel problems require multilevel policy responses. There are no panaceas for addressing the big, persistent challenges of governing for sustainable development. In other words, there is no such thing as a cure-all blueprint policy prescription that will ensure sustainable outcomes. And there is now a substantial body of evidence showing that to address common-pool resource problems effectively often takes more than a single policy actor.[14] It is rare that a single policy actor is in a position to respond effectively to complex social-environmental systems problems, because (1) an actor rarely possesses all the knowledge needed—scientific, operational, or local contextual—to come up with an effective response and then implement it; and (2) the effectiveness of the response will depend on how those affected by the policy response perceive its legitimacy. If a government agency unilaterally designs a policy without active contributions from affected parties, it may not be perceived as a legitimate response, which may complicate enforcement.

These ideas have implications for how governance responses to social-environmental system problems are constructed in that they question the traditionally dominant role of the national government in the governance process. But these ideas also question the ability of local users always to produce effective homegrown responses. One of the insights from the research on multilevel governance is that if a problem spans several levels—in the case of climate change, from individual behavioral choices at the local levels to decision-making processes at the global level—governance efforts to address the problem require contributions from actors that operate at all those levels. This does not mean, however, that a global consensus needs to be reached among all these actors before responses can be implemented. Speaking of multilevel approaches to address climate change, Elinor Ostrom argued that "simply recommending a single governmental unit to solve global collective-action problems—because of global impacts—needs to be seriously rethought."[15] According to Ostrom, the contributions of smaller-scale efforts to reduce emissions need to be recognized. One of the advantages of a multilevel approach to addressing a problem like climate change is the ability to learn from experimental efforts at multiple scales that will benefit all.

Consider Interventions as Experiments to Learn From

Any particular institutional arrangement is an imperfect human construct that will produce partial solutions at best. Because of the large number of uncertainties involved in any social-environmental system— there are so many moving parts that will influence outcomes, and these are not only hard to foresee but also often beyond the governance actors' control—mistakes in the design and implementation of these policy responses are inevitable.

The way London first responded to the increased accumulation of human waste in the streets and back alleys provides a useful illustration. The innovation of flush toilets plus the rules that all houses should use them channeled vast quantities of untreated human waste into the river Thames (the primary source of the city's drinking water). Once it became clear that this was destroying human health, London's policy makers changed the rules, requiring that drinking water be sourced upstream of London and that wastes be discharged downstream of the city. (Of course, this example also shows the limits of governance structures that don't match the physical scale of the problem: growing communities upstream of London could not be compelled by London's governance structures to keep their wastes out of the Thames; the downstream communities that received London's waste had no voice in the governance processes that cleaned London at their expense.)

The lessons learned from policy interventions can be used to improve on the design and implementation features of the policy in an interactive and incremental fashion. Thus, one of the keys to effective governance of complex social-environmental systems, which are full of surprises, is to be able to learn from policy interventions. We will discuss this topic in some depth in the next chapter.

Knowledge Is Power, and Sharing Knowledge Can Be Empowering

The phrase "knowledge is power" suggests that improving one's knowledge through education opens doors and increases one's potential and abilities in life. Students of sustainability, for example, will become more powerful and effective advocates of social change after learning how students, professionals, and individual citizens can make a difference for sustainable development outcomes.

The phrase also suggests that knowledge can be used as a tool of domination and to increase an individual's resources at the expense of others. The sharing of knowledge with fellow citizens can therefore act as a counterweight to existing power asymmetries in society. Sharing knowledge can be empowering, as the new knowledge gained can help level the political playing field, improving the credibility, trustworthiness, and legitimacy of groups that push for social change. Nevertheless, knowledge alone is seldom sufficient to achieve social goals. It can certainly play a large role, but ultimately it is how knowledge is used that matters. Part of that use involves who it empowers in the governance process, and thus how it influences resource allocation, the design of interventions, and learning about outcomes. Knowledge *is* power. And knowledge about sustainable development—how well its goals are being achieved, what invisibilities are misguiding production and consumption processes, which capital assets are being built up or degraded, and which are accessible by whom—can empower those committed to pursuing sustainability. The conditions under which individuals, empowered by this knowledge, are able to contribute to a sustainability transition is the topic of the next chapter.

CHAPTER 5

Linking Knowledge with Action

The conceptual framework for analyzing sustainable development introduced in chapter 2 identifies knowledge as one of the fundamental productive assets that determine people's capacity to extract well-being from the social-environmental systems they inhabit. But knowledge, like other productive assets, needs to not only exist but to be used so that it can promote sustainable development. Linking knowledge with action is harder than it may seem.[1]

The cases introduced in chapter 1 provide some examples of successful, or at least partially successful, efforts to link knowledge with action. Sadly, successes—important and inspirational as they are—remain the exception rather than the rule. More common is the irony of unused discoveries coexisting with unmet needs, with dedicated researchers discovering potentially useful knowledge that never makes it beyond the pages of academic journals. Countless innovations designed by practitioners, engineers, and policy analysts who hope to improve people's lives never get beyond local demonstration projects. Meanwhile, the people, organizations, and other actors who are trying to *do* sustainable development are too often starved for the knowledge and know-how they most need. Clearly, the pursuit of sustainability requires more and better knowledge. But that knowledge needs to be targeted and produced in ways that make it useful. And what is known already could and should be better linked with practice. We believe that forging better linkages of knowledge with action needs to be treated as a priority for scientists, engineers, and the people who fund and promote them.

Why is there a persistent disconnect between knowledge and action in the pursuit of sustainability? What can be done to create better linkages that promote sustainable development? These and related questions are the focus of this chapter. Before plunging into our diagnosis and prescriptions, however, it is important to clarify some key perspectives related to knowledge:

- What kinds of knowledge are potentially useful in the pursuit of sustainability? A causal understanding of how social-environmental systems work, discussed in previous chapters, is certainly pertinent knowledge. So are technologies or policies or other practices that can change those systems. Knowledge on best ways to conduct the monitoring, research, and innovation that produces such usable knowledge is also important.
- Who consumes or uses knowledge for sustainable development? The users we have in mind include all the actors characterized in figure 2.1 and chapter 4: decision makers at all levels, from households to corporations to universities to governments. Also important, however, are opinion leaders, educators, and artists who shape the way we all think about sustainability. Note that sometimes knowledge producers and consumers are the same people, most notably in the case of poor or otherwise marginalized individuals who don't have access to knowledge networks extending much beyond their own upbringing and experience.

With these clarifications in mind, we turn to a discussion of what research in recent years has discovered about linking knowledge with action for sustainability. We focus on six big lessons, which are summarized in box 5.1. We expand on these lessons in the remainder of this chapter.

INFLUENTIAL KNOWLEDGE
IS TRUSTED KNOWLEDGE

Think of a personal decision that really matters to you and that might benefit from use of additional knowledge: For example, what single action should you yourself take if you want to maximally reduce your environmental footprint on the planet? Should you take out a big loan

BOX 5.1. Characteristics of Successful Efforts
to Link Knowledge with Action

1. They are trust-building efforts that reject the notion that experts' declared good intentions ought to be sufficient to propel their ideas into action. Instead, they understand that for actors on the front lines of sustainable development to treat knowledge as sufficiently trustworthy to influence their beliefs or behaviors, they must perceive it to be not only credible (likely to be true) but also salient (relevant to their needs) and legitimate (free from bias and ulterior motives).

2. They are collaborative efforts that reject conventional "pipeline" models of technology transfer and science of policy advice. Instead, they support collaborative efforts in which knowledge producers from multiple disciplines and multiple traditions of practice work closely with multiple sorts of knowledge users to define research priorities, identify relevant evidence, and shape appropriate evaluation criteria.

3. They are systems efforts that reject simplistic assumptions that new discoveries or inventions alone are sufficient to constitute useful knowledge in particular contexts. Instead, they recognize that useful innovation is a complex process requiring joint success at a variety of interactive tasks, including invention, financing, pilot production and adoption, evaluation in use, adjustments to fit particular users, dissemination, and eventual retirement to make way for even better ideas.

4. They are adaptive efforts that reject the quest for certainty in knowledge as a precondition for action. Instead, they acknowledge the inevitability of failure, encourage intelligent risk taking, and seek actively to learn from the often bruising experience of nurturing new ideas and technologies through their encounters with action in the real world.

5. They are political efforts that reject the comforting myth that research and invention are value free. Instead, they accept the reality (noted in the previous chapter) that knowledge is power and take care to assure that their choice of questions to explore, selection of collaborators, and dissemination of results are carried out in ways that promote the well-being of all rather than a privileged few.

6. They embrace boundary work that actively manages the interactions among multiple knowledge producers and users. Rejecting the simplistic view that good will is the only thing needed to link knowledge with action, they acknowledge the tensions inherent in such efforts and invest in people and processes to address them.

to invest in a green startup company founded by a classmate? What field of study should you specialize in if you want to make the greatest possible contribution to sustainable development? Should you take up an offer to help with an exciting sustainability project in a potentially dangerous place? Should you seek out an experimental vaccine for a deadly disease you might encounter on your journey?

Now, suppose that someone aware of one of the choices you are confronting comes up to you and announces that she is an expert on the subject. Will you change your behavior to reflect the self-proclaimed expert's advice? Will you even let her views influence your beliefs? Why?

Dilemmas such as this arise all the time in the context of efforts to promote sustainable development. They raise the question of how knowledge becomes influential. The answer, as always, depends on the particular problem, people, and contexts involved. Our experience, however, suggests that almost always the answers hinge on **trust**.

Trust, as we noted in chapter 2, is an element of social capital, the arrangements that influence how people interact with one another. We use *trust* in the broad sense of belief that people will keep their promises. But what does trust mean when applied to knowledge offered up to help in the pursuit of sustainability? For many of us engaged in the production or dissemination of knowledge—as researchers, inventors, policy analysts, or other innovators—even asking this question can be troublesome. We truly want to be useful, and many of us resent the implication that our good-faith efforts to help are untrustworthy. However, when we, as researchers, recommend that other people take actions on the basis of our expertise, our stakes in the outcome are usually relatively small. We don't have a lot to lose. In contrast, users' stakes might well be a matter of their livelihood or even their life. Moreover, since snake oil was first sold on this planet, producers marketing it to users have claimed expertise as part of their sales pitch. Potential users of knowledge, facing a flood of new ideas and products, therefore have every reason to be skeptical of knowledge-producing experts. Indeed, experience has shown that knowledge is unlikely to influence action unless potential users perceive it to be trustworthy. But where does such trust come from?

TABLE 5.1. DIMENSIONS OF TRUST IN KNOWLEDGE AS PERCEIVED BY POTENTIAL USERS OF KNOWLEDGE

Dimension	Key Question (from potential knowledge users)	Example: Experts recommend a bridge to connect a farming town to a market across the river. Farmers ask:
Saliency	Is it relevant?	Does a bridge address our need to get crops to market?
Credibility	Is it true?	Do these experts know how to build bridges?
Legitimacy	Is it fair/unbiased?	Why are they recommending this bridge? Do they really want to help us, or does someone in their family build bridges for a living?

Research on this question suggests that potential users are more likely to trust new knowledge, and may therefore be willing to act on it, when from their perspective it meets three criteria: **saliency, credibility**, and **legitimacy** (table 5.1).[2] This formulation of the determinants of trust in knowledge intentionally puts the emphasis on perceptions of the user. It is worth noting, however, that users' perceptions can be influenced by the behavior of knowledge producers. In particular, if the community of producers from which a new discovery or invention emerges is itself vocally skeptical of that knowledge, many potential users are likely to take such skepticism into account in shaping their own judgments. This sets up a dynamic in which those seeking to link knowledge with action must pay attention to how their strategies play out in both the producer and user communities, rather than optimizing on one alone.

Saliency Is about Perceptions of Relevance

Do potential users see the expert advice or new technologies they are being offered as bearing on their most important needs? Does the cook see the new low-emissions stove as something that will provide the kind of heat she or he needs to cook the household's meals, or just as an antipollution device? Do travelers see highly efficient mass transit systems as an innovative response to their desire for flexible, on-demand personal transport, or just as another intrusion by the government into their lives? Do householders who worry about blackouts see rooftop solar energy as a reliable source of off-grid power, as a step toward national energy independence, or both? Do college students see locally sourced organic produce as tasty food that will get them through the day, or as an unwelcome admonition about good nutrition?

Our point is not that potential knowledge users—be they corporate leaders or farmers or householders—always know best which new ideas or devices could improve their well-being or society's prospects for sustainable development. They don't. But they have no reason to listen to the producers and purveyors of new knowledge unless the conversation begins with recognition of what users hold to be their greatest hopes, fears, and goals. Commercial innovators know this well; scientists should take a leaf from their book. The story about cookstoves in box 5.2 shows what happens when they don't.

The most likely source for salient knowledge about a problem is often the people who have been grappling with the problem and solutions to it in their daily lives. This may be knowledge of healers about medicinal plants, or of farmers about how to run an irrigation system, or of machine operators about how to improve an automobile production line. In general, such people are both producers and users of

---

BOX 5.2. Unintended Consequences: A Story about Cookstoves in Tanzania

I recall living for a short time during the early 1990s in a Tanzanian community at the foothills of the Usambara Mountains. A well-meaning organization had recently designed and distributed fuel-efficient, smokeless cookstoves. Unlike traditional, open-flame stoves, these cookstoves were closed and had chimneys which directed smoke out of traditional mud-hut homes. They were also fueled by sawdust, rather than wood or charcoal.

The intention was to reduce indoor air pollution dramatically—a source of illness and premature death for women and children who cooked indoors for several hours per day. By using sawdust, the new cookstove was expected to obviate the need for women to walk several miles each day to collect fuel wood, as well as reduce deforestation of the rainforest.

While the technological innovation had great promise for enhancing human health, environmental conservation, and social equity, the community rejected it. Sitting around a bowl-shaped traditional stove, I began to learn from two sisters why the community did not accept the new stove. The women turned their white corncobs over the open flames and explained that roasted corn was a favorite food, and roasting it was a deeply ingrained cultural practice that would not be possible on the new stoves. It provided a time for families to gather, and children took the cobs for nourishment when they were out for extended periods herding animals.

After living with these women for a period of time, I discovered that families relied on smoke in their huts to keep mosquitoes at bay during evening and morning hours, protecting adults and children from malaria. Medicines were too expensive and health clinics too far away to offer an alternative. While burdensome, women's walk for fuel wood was a rare opportunity to connect privately. They deeply valued the opportunity to discuss issues of mutual concern and develop ideas for better supporting one another and their families. Community members also questioned the reliability of the sawdust supply. In the future, once the cookstove project team left, who would provide it, from where, and at what cost?[3]

This anecdote offers a reminder to ground technological creativity and ingenuity in the realities of the complex social-environment systems in which one works. Many organizations now partner effectively with communities throughout developing countries to ensure that design, manufacture, and distribution of cookstoves and other technologies genuinely meet local needs and are suited to the contextual environment. Successful efforts look holistically at challenges and solutions, mapping interrelationships, trade-offs, feedbacks, and unintended consequences as an integral part of the design process.

They rely not solely on scientific expertise but also on human relationship skills—an ability to connect and build trust—to understand the unexpected ways in which cultural, scientific, and environmental factors combine to shape incentives and motivations. In an increasingly complex, interdependent, and unpredictable world, situational awareness and empathy have become even more critical in ensuring that our efforts to advance intergenerational well-being succeed.

knowledge. Some of the most effective efforts to create salient knowledge to inform action on sustainability involve integration of such local, practice-attuned knowledge with contributions from outside experts. This was evident in the case of Nepali irrigation works, where ineffective interventions by outside engineers gave way to more useful problem solving when solutions were designed and implemented by the local builders of irrigation works.

Knowledge can also become salient when knowledge producers recognize the knowledge needs that users say are most salient and adjust research agendas to supply it—even if this doesn't involve cutting-edge work publishable in the most scholarly journals. In the Yaqui Valley case, the researchers avoided comparisons of extreme treatment options (an approach that would maximize differences in results and potentially lead to more complete process understanding) and instead tested options that farmers could actually pursue; in doing so, they still made discoveries but were better attuned and responsive to the community's priorities.

We heard a similar story from one of our colleagues who worked on community response to climate change. He and other researchers were trying to get a coastal community to pay attention to its assessments of the growing risk of storm surge flooding. After much frustration on both sides, it turned out that the community was most immediately concerned with how the culverts under its roads should be resized to reduce the frequency of washing out with the increasingly intense storms being experienced by the region. Once the researchers addressed this mundane (to them) but pressing and salient (to the community) need, the dialogue expanded to cover the coastal flooding issues that had initially attracted the researchers.

More broadly, salience turns out to be highly dependent on how issues are framed and problems are defined. For example, consider the framing of fossil-fuel energy versus renewable energy as critical only in the context of global climate change. An alternative framing would highlight the broad range of impacts among energy choices, some immediately salient to most people, some more distant in space and time. Such a framing would focus on not only the more chronic global damages (climate change and ocean acidification) of emissions from fossil fuels but also more acute and immediate local ones (to health, materials, agriculture, and ecosystems). Thoughtful efforts to develop problem definitions and framings that encompass both the immediate and long-term concerns of potential users increases the likelihood that scientific knowledge will influence actions needed to promote sustainability.

It is important for knowledge producers to realize that their assumptions about what users ought to need may be very different from users' perceptions of what they most want. Reorienting knowledge production efforts to assure that they are (and appear to be) salient to the most intensely felt needs of potential users is an essential step to building influential knowledge.

Credibility Is about Perceptions of Truth

Does the potential user have reason to believe that the person or organization providing new knowledge actually knows what he or she is talking about? Will the no-till agricultural practice really reduce vulnerability to drought? Is the new design for a nuclear reactor really "inherently safe"? Will the new vaccine really protect the kids without causing dangerous side effects?

What makes knowledge credible to a prospective user depends on the circumstances. In academia, those circumstances are widely understood: one scholar may be willing to invest in building her next research program on another's new finding only if that finding has been carefully peer-reviewed and published in a reputable journal. But a poor farmer may be less interested in peer review and more concerned about seeing the results of a new seed variety planted in a neighbor's field before adopting it himself.

Societies are not particularly consistent in what passes for credible knowledge claims. Many insist that claims that a new drug is effective and safe are credible only when the drug has passed elaborate and relatively transparent testing procedures. In contrast, for many societies, the credibility of claims that a new processed food is "good for you" rests largely on the personal judgment of the individual user, ostensibly informed by the "fine print" on the package. For the purpose of linking knowledge to action, credibility must be constructed for particular users in particular contexts using standards that work for them.

At least two related pitfalls complicate efforts to grapple with the credibility of knowledge for sustainable development. The first is the assumption that credibility is *all* that matters in linking knowledge to practice. University-trained researchers may assume that their own best practices for producing credible knowledge (e.g., statistical significance, replicability) are appropriate everywhere, when users may give other criteria more weight, such as demonstrated success in mitigating specific local problems. In turn, local inventions or novel practices that have a real potential for contributing to sustainable development may be shunted aside by the powers that be for being "unscientific." (This is not a new problem. In the London case, elaborated on in appendix A, we noted the Royal Society's rejection of successful "foreign" experience with new treatments for smallpox simply because their scientists knew of no mechanism by which it could work.)

The second pitfall involves the flip side of the first. In the absence of critical criteria to establish the credibility of knowledge claims, mere assertions pile up and can clog the system and forestall progress. The London case again provides an example. There the widely accepted but not critically evaluated "miasma" theory of disease stood in the way of reformers' claims that it was the contamination of water, not its stench, that was the cause of cholera and that needed to be the focus of treatment. Other false knowledge may follow from the good intentions of researchers who hope to discover things to promote sustainability—for example, better cookstoves, or nuclear reactors—but fail to evaluate carefully whether intention and actual performance match. And, more rarely, real progress can be delayed by the self-interested activities of those who knowingly push misinformation for their own profit. The growing popularity of randomized controlled trials (RTCs) for development projects is a response to these risks. RTCs have proven useful for example, in sorting out the claims being made about the scores of small-scale water purification technologies that emerged and were clamoring for support in response to the Millennium Development Goal of halving the proportion of the population without access to safe drinking water.[4] But such trials are not a panacea, and their utility in evaluating the credibility of larger-scale strategic interventions is limited.

Most researchers acknowledge that procedures to establish credibility are essential for knowledge to influence actions. That credibility is ultimately a function of how the knowledge and the people and processes producing it are perceived by particular users. Repairing a lack of credibility therefore requires looking at how knowledge is produced from the perspective of particular users, understanding what evidence and arguments pertaining to credibility would be convincing to them, and working with them to produce what they judge to be adequate evidence and arguments.

Legitimacy Is about Perceptions of Fairness, Lack of Bias, and Respect

Do potential users see knowledge producers as engaged in a sincere effort to help them, in ways that honors their values and beliefs? Or do the experts come across as paternalistic know-it-alls who have predetermined ideas of the "right" way of doing things? Is the extension agent who warns about the imminent arrival of a pest outbreak really trying to help the farmer, or might he (also) be on the payroll of the firm that sells insecticides?

Legitimacy is the most subtle of the dimensions of trust in knowledge. Scientists like to think of themselves as unbiased. Yet, even when picking which problems to study, they often take sides in an unequal struggle in which some groups have more power (political or financial) to frame the agenda than others. Two of us, for example, worked on climate change during a period in the mid-1980 to '90s when the U.S. government began to support research on the physics of climate change but not studies of social responses and vulnerabilities to such changes. Even today, far more scientists study the impact of deforestation on hydropower than the impact of hydropower development on the forest users it displaces. The result of such bias in the selection of problems means that research tends to shed uneven light—and sometimes the most light is focused only on the topics those with money and power want to have illuminated. Knowledge producers who want their findings taken seriously by all stakeholders need to demonstrate their evenhandedness in the problems they choose to study and how they study them.

Even more intractable than fairness are questions of bias. The sad fact is that many knowledge purveyors are seen by users as selling something. Most often, the "something" in play is there to help the seller, not just the user. It may simply be the pet idea of a scientist or advocate who honestly would like to help advance sustainable development but who isn't thinking carefully about those who will be doing the developing. And sometimes, at the extreme, the "selling" is literal, as potential users are bombarded with expert advice that tells them how to improve their lives but that ultimately ends up requiring them to pay the provider for the improvements. That is fine as long as the transaction is one for which the user can assess whether the value of improvements is worth the price; however, it is problematic when the user finds that she can't stop paying for "improvements" that turn out not to be worth the price.

Given that knowledge producers and providers often have multiple agendas, it is entirely sensible of prospective users to require evidence that knowledge producers are on their side, or at least neutral, before trusting their advice. In general, such assurances emerge from long histories of working together, during which knowledge producers establish their legitimacy through their ongoing respect and concern for potential users. The stratospheric ozone case provides an excellent example of this relationship. The participation in the scientific assessments of the problem by scholars from multiple countries, and from business as well as academia and government labs, provided assurance to treaty negotiators that the advice they were getting was not serving the interests of particular corporate or state actors.

"Shorter" pathways toward legitimacy also exist, which is fortunate for knowledge producers just starting out on a new problem. Frequently, these involve teaming up with someone who is already recognized as fair and unbiased by the relevant user community, as illustrated in our Yaqui case study by the Stanford group's teaming up with excellent scientists who were also locally knowledgeable and respected friends of the farming community.

The essential points are that without legitimacy, knowledge is unlikely to be trusted. The legitimacy of the knowledge that is produced cannot be assumed by researchers but must be actively sought out and earned.

BARRIERS AND BRIDGES

We have argued that new knowledge is more likely to be trusted and thus acted upon by users when both it and the processes that produced it are perceived to be salient, credible, and legitimate by those users. We turn next to the questions of what stands in the way of producing such influential knowledge, and how those barriers can be overcome.

A Collaborative Enterprise

A common barrier to linking knowledge with action has been termed the "mutual incomprehension" of producers and users. We touched on this barrier in the introduction to this chapter. Scientists and engineers have radically different notions than do practitioners and decision makers of the most important problems to be solved, the available options for solving them, and the criteria by which the suitability of alternatives can be reliably evaluated.

This mutual incomprehension goes unrecognized in "pipeline" or "technology transfer" approaches to expert advice. In such approaches, questions of needs, options, and evaluation criteria typically are all specified (and assumed) by knowledge producers. The resulting answers that emerge from research are then "tossed over the fence" to decision makers and other frontline actors, with hopes and expectations they will be adopted. The initial efforts of the international community to bring modern engineering solutions to the irrigation systems of Nepal reflected this approach; the development of many new technologies today does likewise.

Such pipeline efforts have had some successes, especially when run by scientists and engineers with a deep and long-term interest in the well-being of the users they are trying to help, and a resulting understanding of what is needed to achieve it. More often, however, the pipeline model, in which experts pass information to users and assumed users, founders on one or more of the criteria for trusted knowledge production outlined earlier. Producers may misperceive the needs of users and fail on the salience criterion, as in the case of the improved cookstoves described earlier. Or they may fail to understand that statistical significance—their criterion for credibility—may be less important than a demonstration of success on a neighbor's fields in establishing the credibility of a new technology or practice. Finally, scientists and engineers often fail to comprehend that their own good intentions are not sufficient to establish the legitimacy of the advice they furnish.

Successful efforts to link knowledge with action for sustainability deal with the mutual incomprehension challenge by becoming collaborative enterprises. Knowledge producers and users work together to identify needs, to design possible options for meeting those needs, and to evaluate the performance of alternative options against mutually agreed upon criteria. Such a collaborative endeavor is where the Yaqui Valley project eventually ended up, with local farmers, farmer credit unions, national and international research centers, and local and foreign scientists all working together to ask questions and test options. Such collaborative enterprises not only make the resulting products—such as agreements on plans and interventions—more trusted but also increase the participants' ownership of these products and their willingness to be influenced by them.

A Systems Enterprise

A second common barrier to linking knowledge with action is the "fragmentation" of the knowledge production enterprise. The collaborative approach just discussed can assure that the perspectives of knowledge producers and users are aligned. Unfortunately, however, the production of that knowledge often requires the contributions of many different pieces of knowledge, most often produced by various scientists or engineers. Without integration of multiple producers' activities, the overall effort can fail to meet the needs of the users. For example, the introduction of the flush toilet in London in the nineteenth century was positively dangerous without a sewer system to take the wastes away from the city's water supply. The new nitrogen-responsive crops produced by the Green Revolution discussed in our Yaqui Valley case would have been worthless without the earlier innovation of the Haber-Bosch process for producing ammonia from the nitrogen in the air and hydrogen, to make cheap nitrogen fertilizer. The utility of today's photovoltaic systems for generating electricity is limited by the lack of cheap and easily recharged batteries to provide storage.

These and other examples suggest that making individual discoveries and producing inventions attuned to user needs is only one part of a much more complex system of innovation. Unless all the parts of that system are in place and work together reasonably well, even research with the highest potential for advancing sustainable development will remain little more than another laboratory curiosity or library article. Achieving the needed integration is hard. Most university-based researchers resist taking orders to work on the parts most needed for a useful product. Government agencies and labs remain stovepiped. And even groups like the international agricultural research centers, established for the explicit purpose of helping improve agricultural technologies at large scale, find themselves constantly torn between the desire of many of their scientists to pursue research frontiers, their funders to support particular ideologies in development paths, and the desperate need for more mundane work in adapting existing and new knowledge to local conditions at the field level around the world.

How can systems of innovation for sustainable development be improved? One promising approach is to give more emphasis to use- or solution-driven approaches to innovation. This does not mean doing away with efforts that begin with discovery and invention of new knowledge and then search for useful applications. It does mean complementing such efforts with others anchored in the problem to be solved, and working backward to discover or invent the knowledge needed to solve it in ways relevant (salient) to users.

A good example is provided by the Regional Integrated Sciences and Assessment (RISA) program set up by the U.S. National Oceanic and Atmospheric Administration (NOAA).[5] NOAA's basic research capabilities in climate change and variability are certainly the anchor for the program, but instead of waiting for users to ask for climate forecasts, NOAA set up RISA offices in a number of regions around the United States to work with local decision makers to identify specific climate-related problems faced by those regions. RISA project teams were assembled and tasked with providing information, tools, and advice to support decision makers in crafting solutions to their particular problems. As a result, improved management actions were developed for dealing with coastal emergencies, forest fires, water resources, and other acute local problems.

Such use-driven innovation systems for sustainability do seem to require a central manager with the resources to assure that all stages of the innovation system are supported. But that role can be played not only by governmental organizations such as NOAA through its RISA program but—if adequately funded—by university-based groups,

NGOs, and public-private partnerships as well. The key to linking knowledge with action is to treat it as the systems innovation problem that it surely is, rather than emphasizing which organization does the work.

An Adaptive Enterprise

A third barrier to linking knowledge with action is that knowledge producers too often aim for one-size-fits-all, static solutions for a heterogeneous and dynamic world. "Adaptation" occupies a disproportionate share of the governance framework summarized in figure 4.1 for a good reason: the pursuit of sustainability is a wrestling match with the unknown, in which ignorance, error, surprise, and dynamically changing rules of the game are ubiquitous. Successful efforts to link knowledge with action function dynamically, as an adaptive enterprise that facilitates learning from encounters with a spatially varied and temporally changing world.

Successful adaptation is rarer than it should be. In our experience, several barriers are especially problematic in the pursuit of sustainability. The first is that it generally requires admission of error. Many people and organizations would rather not admit that they have been wrong, partly for personal reasons and partly for fear of being penalized in one way or another. The result is to deny themselves and the rest of us the opportunity to learn from past mistakes and adapt our understanding accordingly. Everyone will have examples of this pathology. The London case we introduced in chapter 1 included several examples, of which the most extreme is probably the reluctance of the scientific and medical establishment to reject the wrong-headed "miasma" theory of disease. Our own experience includes "sticky" cookstove designs that kept being pushed on rural populations even when it became clear that they didn't work (see box 5.2), and natural resource inventory protocols for fish and forests that were kept in place—misleading decision makers—long after revelation of systemic measurement biases undermined their credibility. The list is endless.

Another impediment to adaptation in the interactions between knowledge and action is the lack of forums for learning from one another's experience. Most sustainable development problems being faced somewhere today have previously been faced somewhere else. There are plenty of forums for learning about successful experiences—journal articles, Web sites, best-practice awards, and the like. There are vastly fewer forums for presenting instructive autopsies of efforts gone wrong. This situation reflects not only the personal and organizational reluctance to admit error noted earlier but perhaps also a fundamental misunderstanding of how risky and uncertain a business sustainable development is, and how important an active error-embracing attitude is for those who would promote it.

What practical steps can be taken to promote knowledge for sustainability as adaptive management? In chapter 4 we discussed adaptive management as a general approach to governance of social-environmental systems; clearly, such approaches need to have a strong commitment to learning from error. Enabling such learning requires "safe spaces" in which partners in collaborative efforts to link knowledge with action for sustainability goals can come together in an environment that supports rather than punishes failure. We have found that universities and international research centers have special opportunities for providing such safe spaces. Much more so than most government or private labs, they can provide a supportive environment for tackling politically sensitive questions, protecting scientists from inappropriate pressures, and devising evaluation mechanisms that reward risk taking. Such settings are hard to maintain, but they matter. And if you want to pursue knowledge for sustainable development, you should find one in which to work.

A second and related practical step is to develop metrics and evaluation procedures that reward learning rather than success, and punish those who hide error while rewarding those who embrace it. A colleague of ours who ran an industrial research lab told us that he pushed his scientists and engineers to take more risks if their reported success rates were higher than 30%. We don't know of a single university department or research funding agency where a 70% failure rate would be something to brag about in a promotion file or grant application. Increasingly, however, some research funders—from the MacArthur Foundation to the Defense Advanced Research Projects Agency (DARPA) to the sponsors of the X-Prize—are positioning themselves squarely in the business of supporting more risk taking and innovation. So are contemporary organizations built around "fail-fast" startup cultures and

the worlds of action and policy making"[6] with practice-based and other forms of knowledge. The central idea of boundary work is that tensions arise at the interface between actors with different views of what constitutes trusted knowledge, and those tensions must be managed effectively if the potential benefits of knowledge are to be realized by society. Too little permeability of the boundaries separating science from the world of action means too little learning from or contribution to practice. Dissolving the boundary entirely, however, risks science being hopelessly politicized, and politics and politicians being viewed as mere mouthpieces of an arrogant and unreliable technocracy. Active boundary work is therefore required to manage effectively the interfaces among various stakeholders engaged in harnessing knowledge to promote action.

In the case of sustainable development, it is clear that an ongoing connection is necessary between expert/researcher communities and the broad range of stakeholders making decisions; harnessing knowledge to promote action for sustainability requires it. However, this turns out to be a very significant challenge for both communities. Working at the interface between science and policy—and more broadly, seeking to mediate activities between knowledge and action to lead to better decision making—is often outside the goals of research institutions and something that few individuals have been trained to accomplish. Different communities have different cultures, different languages, different kinds of evidence, and when multiple stakeholder groups and multiple kinds of science and technology experts are involved as well, the cross-cultural challenges are even greater. These groups characteristically have different views of what constitutes trustworthy knowledge. Bridging those various boundaries is nonetheless essential, and thus boundary work at the interface between communities with different cultures is at the heart of a sustainability transition.

Figure 5.1 illustrates the position of boundary work in the midst of one international effort to devise more sustainable ways of managing human use of the forest-agriculture interface in the humid tropics. In this case the crucial boundary role was played by the Alternatives to Slash and Burn (ASB) program[7] of the Consultative Group on International Agricultural Research, the consortium that helped bring about the Green Revolution. As it was organized in the wake of the 1992

United Nations Conference on Environment and Development (UNCED), ASB faced enormous challenges of linking knowledge with action. First, the urge to act was far out ahead of a trusted knowledge base on which to base effective action. Activists and international program entrepreneurs concerned with the problem of tropical deforestation were so (wrongly) convinced that "slash and burn" agriculture was the cause that they (misleadingly) embedded their diagnosis in the title of the program meant to investigate the problem. Second, just as in our Nepal case, a great deal of the needed knowledge about sustainable and unsustainable uses of the forest/agriculture interface was embedded in local practices distributed around the world. But there was no mechanism—not even a language—for comparing and learning from those local practices. Third, relevant scholarship was distributed across disciplines—soil science, tropical ecology, economics, geography, anthropology, and the like—that rarely worked together on tropical land use issues. Finally, land use on the tropical forest margins was an intensely political issue, with issues of national sovereignty, economic development, environmental protection, and rights of indigenous peoples all intertwined. How was a (misnamed) program ever to link these disparate forms of knowledge together and harness them to inform action in a highly politicized arena?

The answer turned out to require transforming ASB into the "boundary organization" portrayed in figure 5.1: an organization that came to see its central mission as linking the various players involved in the problem in ways that enhanced the saliency, credibility, and legitimacy of the knowledge it produced and mobilized. It did so by grounding itself in a set of local study sites throughout the humid tropics, incorporating in their leadership team local leaders trusted in both the science and policy communities, facilitating research based in local realities but integrated through the development of common field protocols, and remorselessly engaging outsiders from both knowledge and action communities to help them reflect on how they were doing. Through a decade of work, ASB became a trusted source of knowledge, both locally in its study regions and globally as a central participant in the Millennium Ecosystem Assessment. Its work has not solved the problem of sustainable land use at the forest/agriculture interface in the humid tropics; but through its commitment to boundary work, ASB has signifi-

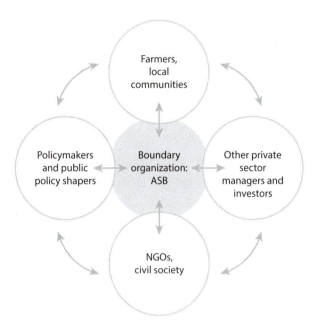

FIGURE 5.1. Boundary work spans different cultures and different communities. (Adapted from S. Liu. 2004. "Strategic Typology of Impact Pathways for Natural Resource Management: A Case Study of the Alternatives to Slash-and-Burn Programme." Report. © 2004 by the ASB Partnership for the Tropical Forest Margins at the World Agroforestry Centre. Reprinted with permission)

cantly improved the linkage between knowledge and action in efforts grappling with that problem.[8]

What can be done more generally to help promote the effective boundary work needed for linking knowledge with action in support of sustainable development? There are many examples of such work carried out by organizations that have been purposefully and explicitly set up to do so.[9] In many of these cases, the organizations involved have within them individuals who view the boundary-work role as a critical part of their job. Outsiders can also more informally play this critical role. Many of the cases also involved the production of what the field has dubbed "boundary objects," collaborative creations such as reports, models, maps, apps, or standards that are meaningful to the combined community and are used to communicate with and engage multiple stakeholders in decision making. The International Ozone Trends Panel

FIGURE 5.2. A boundary object. The Batang Toru watershed in North Sumatra province, Indonesia, still contains approximately 110,000 hectares of forest that harbors a genetically unique orangutan population. Proposals to designate the area as National Park would imply that people would be required to move out. Alternative conservation strategies were proposed to respect and enhance the stability of the agriculture/agroforest/forest gradient. This illustration was developed through revisions of drafts between an ICRAF/Winrock team and villagers as a visual statement that agroforests with planted (rubber) as well as naturally established fruit trees form a buffer between the village and the remaining forest on the hills. The orangutans shown use the agroforest as part of their habitat and are not seen as a threat. The illustration, printed as a poster, served as a boundary object in negotiations among villagers, local government, and conservation authorities, supporting the gradient perspective on integrating conservation and development. (Tata M. H., et al. 2010. *Human Livelihoods, Ecosystem Services and the Habitat of the Sumatran Orangutan: Rapid Assessment in Batang Toru and Tripa*. Project Report No. RP0270–11. World Agroforestry Centre (ICRAF), Bogor, Indonesia. Illustration by Winoyo)

assessment is one such boundary object. Another, developed in a context of tropical forest management similar to that addressed by ASB, is shown in figure 5.2.

Scholarship on what succeeds and what doesn't in boundary work is rapidly expanding. Research suggests that boundary work is more likely to be effective if all relevant stakeholders, including the research community, are accountable for the results. Thus, boundary organizations or individuals that have one foot in the research side and one foot in the decision-makers side of an issue, and are seen as accountable to both, are most effective in the role. Second, such organizations or individuals should be able to tap into the knowledge and know-how and experiences of the multiple stakeholders, integrating them in the development of useful technologies or policies. Such incorporation of many different kinds of knowledge is critical. Finally, scholarship suggests that boundary work that encourages and allows collaborative production of knowledge and boundary objects is more likely to be successful in getting new knowledge to be trusted and successfully used in decision making.

For many of us in academic or other research organizations, working at the boundary can be challenging—it takes time and is all too often not appreciated and valued by our employers. Finding other boundary workers with whom to work is one reasonable solution; university researchers can form close links with action-oriented organizations that will take the lead in fostering collaborations and testing new ideas. Another way of getting academic researchers more user-oriented is to make better use of the classroom as space for exploring collaborative learning opportunities. Some universities have started to reward their professors for teaching service-learning courses. For example, a course on environmental policy, civil engineering, or field ecology may be designed in partnership with a local government agency, NGO, or the university's building operations so that the outcome of the course is not just knowledge that the students value but also knowledge that benefits these partners. In other research organizations, action arms are being developed to provide some of the linking functions. Much more such work in other contexts is needed to fully develop the potential contribution of knowledge to action in the pursuit of sustainability.

CHAPTER 6

Next Steps: Contributing to a Sustainability Transition

Every day, we see creative efforts to address persistent and difficult sustainability problems. Local to global communities are rallying around shared visions for sustainability goals—for livable cities, use of resources, food production, climate mitigation and adaptation, and other issues of well-being. In universities and research centers, new knowledge, tools, and approaches are being developed to meet sustainability goals, and they are increasingly being designed, tested, and employed by decision makers in partnership with research communities and other stakeholders. Future leaders are being trained to understand connections, analyze interactions, engage in governance and lead change locally and also at a scale that can make a difference globally and for the long term.

Good things are happening, but the global community has a long way to go, and there are some daunting tasks ahead. Sober assessments of social-environmental systems show that these systems are in real crisis. The community's efforts to address sustainability challenges need to be greatly expanded and intensified, and all of us need to engage. The world is in a transition period, but that transition to sustainability must progress at a much faster pace in the future than it has in the past.[1]

There are many ways each of us can contribute to this transition to sustainability. Those of us in academia and research institutions (students, faculty, researchers, and staff) are in the knowledge business, and we can help both by creating new knowledge that is designed to be useful and used, and by training the next generation of leaders who will engage in the transition. Others—working in multinational to family businesses; NGOs; local, state, and national governments; and the military—have huge roles to play, often being most responsible for articulating needs and using knowledge in decision making and action. And much of the work is being done, as always, by decision makers on the front lines: citizens who are grappling every day with the challenge of making decisions that make sense for themselves, their neighbors, and future generations. Everyone has a role to play in the transition toward sustainability.

The perspectives that we've shared in this book are meant to help individuals, and especially scientists, be *agents of change* in the sustainability transition. We've presented a framework for relating the sustainability goal of intergenerational well-being to five underlying assets that ultimately determine society's ability to meet those goals. We argued for the importance of paying attention to and understanding the complex interactions in social-environmental systems in which the pursuit of sustainability is played out, and being aware of how one's own and others' expertise can contribute to that understanding. We suggested that an appreciation is needed of the governance processes through which agreements and decisions are made, allowing communities to achieve together what cannot be achieved alone. And finally, we've drawn attention to the ways that useful knowledge, tools, and approaches can be developed in collaboration with decision makers and other stakeholders at all levels.

While these general perspectives are meant to be a guide to help individuals conceptualize and characterize and address the particular sustainability issues they encounter in their personal and professional lives, they are not, in themselves, enough. They are just part of a larger toolbox needed to make headway toward sustainability. Each individual also has in his or her own toolbox a set of specific skills and knowledge, accumulated from experience and education, and a mindset that prepares and allows him or her to contribute in significant ways. A quick read of each of our case studies in appendix A brings home the importance of different motivations, different subject matter knowledge, and different kinds of analytical, reasoning, and technical skills, but the exact mix will be different in every social-environmental system context and for every complex challenge. But it would be impossible, in this short book, to specify all appropriate knowledge and skills—all the things that need to be learned. We hope our readers will be motivated to develop those knowledge bases and skills through courses and experiences throughout their careers. We do, however, conclude this book with some reflection on the mindsets of sustainability leaders who are effective agents of change.

MINDSETS OF SUSTAINABILITY LEADERS

What are the characteristics of leaders who can catalyze progress toward sustainability goals? In our roles as scientists and educators, we have had the opportunity to ask this question of many different kinds of leaders—corporate, educational, non-governmental, and governmental. In doing so, we have received a remarkably consistent set of opinions and a remarkably clear description of the mindsets that appear to be characteristic of leaders in this realm whether they operate on a local or global stage.

First, sustainability leaders are systems thinkers, and they seek to understand the complex social-environmental systems in which they work, and at larger scales as well. To do so, they must have open minds. As they try to incorporate an understanding of social-environmental systems into decision-making processes, they bring their own multidisciplinary perspectives to bear. Importantly, however, they recognize all that they don't know, and they depend on and continue to grow from others' expertise, and respect others' different ways of knowing and learning. They have the ability to build and rebuild multidisciplinary teams and collaborative groups to accomplish a community's goals. They are inclusive and build respectful, collective, and collaborative processes.

Second, sustainability leaders are deeply reflective and adaptive thinkers. Given the complexity of social-environmental systems, they recognize that their own mental models and ways of thinking are unlikely to be how the world really works; they seek out alternative models, push to find better models, and make corrections in their paths. With their communities, they cocreate visions for a sustainable future, but they pursue those visions with a recognition of reality. They are comfortable with complexity and ambiguity, they are iterative, and they continuously learn and adapt.

Third, sustainability leaders are self-aware, with a deep empathy and compassion for the well-being of others. They deeply believe that the sustainability challenge is, at its heart, about inclusive well-being of people, around the world and over time.

Last in our short list is creativity and innovation for change *at scale*: sustainability leaders have a collaborative and strategic drive to create and implement new pathways that ultimately transform social-environmental systems through a process of change that can cascade across generations and can make a difference for human well-being locally and across the globe. Such change takes time, and effective sustainability leaders pursue it with patience. They are innovative in the truest sense of the word, and they see themselves as a creative force for achieving sustainability goals that will make a difference not just for themselves and their communities and companies but for others around the world and in future generations.

FROM INTENTION TO ACTION

With the perspectives summarized in this book, an awareness of the mindsets of sustainability leaders, and one's own particular set of skills and knowledge, what are the next steps one should take—the steps from intention to action? Each one of us has to construct that sequence for ourselves, and there is most definitely no one right way to do it. However, it may be useful to share some insights from individuals who have begun the journey.

Maria Foronda: An Environmental Activist in Peru

Let's start with Maria Foronda, who is an activist working for an NGO called *Natura*, based in her hometown of Chimbote in Peru.[2] Maria Foronda, who was trained as a sociologist, has spent most of her life working to improve the health of the people who depend on the coastal ecosystem in Chimbote. One of the big threats to this social-environmental system is pollution from the fishmeal industry—the industry that turns fishing by-products into animal feed, fertilizers, and preservatives among other products. Many of the fishmeal factories are located near residential areas and use old and inefficient technology that fails to remove much of the waste from the production process before dumping concentrated fish remains (including oils, soluble proteins, and boiling hot water) into the environment. The pollution has caused major health problems for the local population, including fungal skin diseases and respiratory illnesses. Even the outbreak of cholera in Peru in 1991 has been linked to pollution from that industry. And off the coast from the factories, marine biologists have documented expanding "dead zones," absent of any life. The social-environmental system of the entire coastal region was (and to a large extent still is) under severe stress.

This was the environment that Maria grew up in. Moved by the injustices—both to the local people and their natural environment— Maria decided to take up the fight to defend the well-being of her community. She decided to study sociology to become a social worker, which she felt would be an effective way of helping her people address the problems they were facing. She completed her master's degree in Mexico and then moved back to Chimbote to volunteer as a social worker, and her career as a community activist began. She helped found *Natura*, an NGO with a mission to promote the well-being of people and the ecosystems they depend on. Through *Natura*, she organized fact-finding missions with local residents to find out where the pollution came from, what it consisted of, and what harm it caused. With the help of local researchers, they documented the extent of the negative externalities caused by the fishmeal production. She mobilized broad collective action among local community members to protest against fishmeal companies and put pressure on the factories to reduce the pollution.

Her actions did not go unnoticed by the powerful fishmeal industry or the government officials supporting it. In 1994, they accused her of conspiring with the guerilla organization *El Sendero Luminoso*, and a court sentenced her and her husband to twenty years in prison. But as a result of an international NGO campaign that put political pressure on the Peruvian president, they were released after thirteen months. During her time in prison, Maria not only gained increased international recognition and support for her work to fight the contamination problem in coastal Peru, but she also realized that a new style of activism was needed to address the pollution problem more effectively. She concluded that her early strategy of activism—characterized mostly by confrontational protest and accusations directed toward the powerful industries—was no longer going to work. Inspired by the work of NGOs in Europe and North America that had been successful at promoting social responsibility in the private sector through negotiation and engagement in joint planning and collaborative work with private companies, she was convinced that the time was ripe also for *Natura* to pursue such a strategy.

The accomplishments resulting from this new approach are remarkable. Contamination from the industries was reduced significantly, as eight of the major fishmeal companies decided to invest in cleaner technologies. Less contamination and improved knowledge among local people about the pollution also led to fewer health problems (although they were far from eliminated). What exactly did Maria and her organization do to convince these companies to change their ways?

In an interview with grist.org, she explained that through her team's engagement with the companies, it became clear that it is "profitable to invest in clean technology. It reduces the losses in their entire production process, they save on raw materials, improve their productivity, lower their costs—and they don't harm the environment or the health of the communities."[3] Maria and her colleagues also pointed out that international markets are changing and are increasingly scrutinizing producers' environmental records. Through several years of persistent work and a unique set of partnerships between community groups, fishmeal producers, and the government, the implementation of more sustainable business practices for fishmeal production was under way.

Ray Anderson and Interface Carpet

Let's tell another story, this time about Ray Anderson, the founder of Interface Carpet in 1973. Mr. Anderson passed away in 2011, but his legacy lives on in his company's mission and culture. At first, the company did business like many other industrial manufacturers—sending all its waste to landfills, employing over 900 toxic chemicals in its manufacturing process, and simply not considering very intensely its relationships to the world outside of its profit goals.[4] In 1994, however, after reading Paul Hawken's book *The Ecology of Commerce*, Ray became a sustainability leader. He analyzed the social-environmental system that included his business and identified several major areas in which it could make progress. Crucially, he enlisted the talents of all his employees to help in the effort. With internal programs like Quality Using Employee Suggestions and Teamwork (QUEST), Ray fostered multifaceted teams to work creatively through problems. Anyone from a factory floor worker to a top manager could be on the same team, thinking through particular challenges (such as designing a new water-efficient fiber-dyeing system). In this way, he and his employees pushed Interface to reach toward goals that included eliminating waste, creating only benign emissions, reclaiming and recycling its products, and redesigning the carpet business to focus on service instead of production.[5]

Since 1995, Interface has made significant progress toward its sustainability vision. By fostering an inclusive corporate culture that encourages any employee to make suggestions and to problem-solve, Interface has succeeded in reducing waste by 60% and reducing its greenhouse gas emissions by 78%. Carpet looms were redesigned, water use was reduced by over a million gallons per year, and a system for separating the front and back of carpet tiles—and recycling both—was invented. Everything possible within the company's purview is reused in some way—even the restrooms' paper towels are burned for fuel.[6] Though it has long used recycled plastic bottles in its carpets, in 2011 Interface began sourcing some of its nylon from discarded fishing nets in the Philippines, helping eliminate a waste problem there while increasing its sources of recycled materials.[7] Although Interface's energy- and water-saving initiatives have saved the company money ($40 million in 2005), using recycled materials still costs more than manufacturing brand-new carpets. Nevertheless, in total over the past fifteen years Interface's sales have climbed 66%, and profits have risen 28%. Financially speaking, the company is certainly sustainable.[8]

Though Interface is aiming for "zero impact" by 2020, it still faces a number of challenges. The company's goals are to create materials that last and yet eventually degrade harmlessly back into the earth. Technologically, that's challenging, but industry and social standards and rules (e.g., for flame-retardant properties) also create barriers to progress.

Lack of more-sustainable suppliers (e.g., for fabric dyes) is another challenge that Interface is working on. Overcoming such barriers will allow change at scale, and today, many more manufacturers are moving in the direction of a "circular economy."[9]

The Yaqui Valley Case Study Research Team

As our final example, let's return to the Yaqui Valley case study and reflect on how the researchers involved in it moved from intention to action at scale.

Prior to that collaborative effort, we scientist leaders had worked quite independently in our own disciplinary areas—including biogeochemical cycling and global environmental change, food production and poverty alleviation, and agronomic innovations for the developing world. We brought these interests together with the intention of making a difference for sustainable development in the Yaqui Valley, but we also wanted to influence agriculture and its impact on human well-being and environmental change at the global scale. We wanted to make a difference locally and globally.

At the very beginning, we recognized the importance of understanding the whole social-environmental system, not just parts of it. But we also realized that we wouldn't be successful in that understanding if we just tried to "glue" our different research efforts together. Rather, we needed to ask research questions together, and as a result, the research we did together was different from what any one of us would have done alone. With respect and trust for each other's expertise, we collaborated as a team. We were open-minded and willing to learn from one another and from our partners in the valley; we engaged in collective, collaborative leadership of this fifteen-year research effort.

We also recognized the critical importance of engaging with decision makers. Ivan Ortiz-Monasterio, our agronomist colleague who lives in the Yaqui Valley, was one of our key boundary individuals; he is known well by the farmers and is also a trusted member of the research community. When necessary, he reminded the research team that we were talking not just about research outcomes but people's livelihoods and well-being. He gave us the credibility and legitimacy we needed to engage within the social-environmental system, and ultimately helped shift the dialog in the valley from one focused on increasing agricultural yields to one that is now at least including sustainability perspectives.

It took time and work and learning from failure to understand decision making in the region and to recognize that critical decisions were being made not just by the farmers but by their credit unions and, of course, the policy makers at the state and federal level—many of whom were part of the agricultural community. As we gained understanding of the governance systems there, we learned how challenging power dynamics can be. With nonagricultural voices only weakly heard, and many environmental issues invisible to the farming community, it took time to change the conversation. Interestingly, the research we did there brought previously invisible issues to light and was in part responsible for changing the conversation. Some of that research would not have been possible if not for international sources of funding (such as from CIMMYT, the Packard Foundation, other U.S.-based foundations and agencies, and Stanford University). Knowledge is power, and money in support of research, or lack of it, can determine what is known and thus what can be used in dialog and decision making.

Finally, we cared about "scaling"—we wanted to make sure that the solutions that worked here could be scaled to other regions. We developed models and technologies and approaches that would be useful not just in that place and time but in many others. We published and shared those approaches, and through Dr. Ortiz-Monasterio as well as other members of our team, we collaborated with international research organizations and efforts that could help carry those approaches beyond the Yaqui Valley. The work is not over, and never will be, but there is no doubt that progress is being made in a transition to sustainability.

***

The Yaqui team and the other leaders discussed in this section learned an important truth: sustainability is not something to be achieved but is instead something to be worked toward! The world keeps changing, new challenges continue to emerge and change the game, and some seemingly smart things don't work. But each step forward matters, and the sooner the better. Good intentions aren't enough; each of us needs to take the next steps toward sustainability.

Launching a new company focused on "green" cleaning detergents, superefficient fertilizer application technologies, solar systems, biotechnology development of new materials, or information technologies that make it easy to be energy efficient are the kinds of actions that get attention. But personal actions like driving less and eating less meat, composting and recycling, or marching for climate action in Washington or New York are also critical. Science and technology are important, but we, the authors of this book, fear that by focusing too much on the potential benefits of more knowledge and innovation we run the risk of becoming complicit in the continued depletion of our assets, ultimately endangering the lives and well-being of people today and in the future, as well as many of the other species with whom we share the planet. As Amartya Sen noted, "informed agitation" is needed. He writes:

> The challenges that we face in so many different fields—from population growth to the explosion of material consumption and proliferation of wastage—call for something much more than technically cunning recommendations. Scientific analysis, which is crucially important as a first step . . . should also lead the way to much broader exchanges, deliberations and informed agitations.[10]

FROM INTENTION TO ACTION
IN UNIVERSITY SETTINGS

Given our personal roles as faculty in three different universities, and because many of our readers are likely to be students and teachers, we find it compelling to reflect on necessary steps from intention to action in the academic world.

First, it is clear that just about every discipline has a role to play in the pursuit of sustainability. Deep expertise in a given subject is often necessary, but isolated and "siloed" knowledge and expertise can also be limiting. The ability to understand and respect other kinds of knowledge and different ways of knowing, and to recognize how they intersect and interact with one's own, is a central component of most of the effective efforts we know to link knowledge with action for sustainability. Many universities today are attempting to break down silos by creating interdisciplinary departments and schools and cross-university institutes, incentivizing multidisciplinary interactions, and creating opportunities for faculty to carry out research and teaching together. Moreover, cross-disciplinary educational programs focused on sustainability challenges are increasingly being established. Sustainability science is becoming recognized as an emerging multidisciplinary field of study (although not always under that name), in a way very similar to the multidisciplinary agricultural science and health science fields that emerged in the twentieth century.

How can universities best prepare students to work toward sustainability goals? In some universities, students may pursue interdisciplinary degree programs that link physical, biological, social sciences, engineering, and even humanities. Within such programs, obtaining depth of some kinds of knowledge along with breadth is a widely recognized necessity. Other students study in more narrowly focused disciplinary programs but at the same time pursue opportunities for integrating their knowledge with others through multidisciplinary team efforts in the classroom and research lab, and in the real world. Experiential learning that allows students to frame problems jointly with decision makers is a central part of many of these educational processes. Collaboratives of students and decision makers, internships, practicums, team research projects, and public service learning are all mechanisms by which experiential learning has been achieved.

Within academia, involvement in the creation or teaching of sustainability science (by whatever name) can lead to challenging discussions with those colleagues who perceive the "purity" of intellectual pursuits as the only or the most important role of the university. Many of us in the academic research world face a balancing act between our responsibility to contribute to general learning—to discover new things about the world—and our desire to create the knowledge that directly helps design, implement, and test practical solutions for sustainability. There is still, in some universities and funding organizations, a debate about the worthiness of "basic" versus "applied" research; in some places, academic rewards are more heavily balanced toward the former.

We find Donald Stokes's analysis of this debate most useful to us as we try to explain what research looks like in the sustainability science realm. As shown in figure 6.1, research can said to be focused on dis-

Research inspired by...		Considerations of use?	
		No	Yes
Quest for fundamental understanding?	No	'Soaking and poking'	Applied research (Edison)
	Yes	Basic research (Bohr)	**Use-inspired research (Pasteur)**

FIGURE 6.1. Use-inspired research is central for the field of sustainability science. (Stokes, D. E. 1997. *Pasteur's Quadrant: Basic Science and Technological Innovation*. Washington, DC: Brookings Institution Press)

covery, on problem solving, or on both simultaneously. There is no doubt that sustainability science both builds on the discoveries of research done simply for discovery (i.e., "basic" research) and employs the approaches and tools of "applied" science. But we find that the most exciting sustainability science is *centered* in "use-inspired" research—research that simultaneously contributes to discovery and to problem solving and facilitates interaction between these two forms of research. Again, drawing on our personal experience and that of many colleagues, we know that is possible to do both, often at the same time—just as Louis Pasteur did, becoming Stokes's prototype for "use-inspired" research. Sustainability science (encompassing the broadest range of social and biophysical sciences, medical sciences, and engineering) is a new field of such use-inspired research, and it is brimming with opportunity for discovery as well as meaningful contributions to the well-being of humans and the conservation of our planetary life support systems.

Beyond the debate about the kind of science (or knowledge production) done in universities, we often also face concerns and challenges from colleagues and the public around the perceived sacrificing of our roles as scientists for roles as advocates. Academic communities by and large recognize that today there is no such thing as purely objective science. At a minimum, we choose our research areas subjectively, because of personal scientific interest in the topics, but we also are influenced in choosing them by public perceptions, funding opportunities, and our own beliefs and commitments. Nonetheless, internal debates within universities sometimes imply that if research is not disinterested, it can't be good. For those of us academics working on sustainability issues, however, the claim that we should be disinterested researchers working in neutral situations is untenable. Rather, we embrace the idea that our research is important to society and admit that we care how it all turns out! One of our biggest challenges, then, becomes how to stay true to science and to high-quality research: how to focus not on proving that our wishful thinking is right but on rigorously testing our knowledge to figure out where our understanding is lacking, and continuing to learn and explore what is more likely to work. We try to show our students that it's possible, indeed essential, to do committed research and excellent research at the same time.

CONCLUDING THOUGHTS

A debate persists on whether humanity's potential "to make development sustainable" heralded by the Brundtland Commission a generation ago can be realized in fact and in time. Optimists in that debate read the relevant chapters of human history as a series of "just in time" adaptive responses—often using human, knowledge, and manufactured capitals—to address looming environmental scarcities and limits. Pessimists see things differently. Their history lessons focus on the frequency of environmentally mediated social collapse, and they tell stories of how technology can make us more efficient at doing bad things.

How do we, the authors of this book, view the debate? How do we see the pursuit of sustainability playing out? We believe that the world is already embarked on a transition toward sustainability but that progress can and must be accelerated. We believe that efforts to enhance the prospects for sustainability will fare better when they pay attention to the underlying asset stocks that are ultimate determinants of inclusive well-being, including but not restricted to nature and the essential life-support services it can continue to provide if wisely used. We find recent achievements in human development truly inspiring, because they show it is possible to address persistent and difficult problems at a massive scale, even while recognizing interconnectivities and complexities. We look to a time in the not too distant future when improvements in health and opportunities for all people cause the world population to level out or even decline and give everyone an improved capacity to learn and create sustainability solutions. We look forward to the blossoming of technologies, tools, and approaches that help decision makers understand the full dimensions of their decisions. We engage in the training of future leaders who understand connections, care about the future as well as the present, and have the drive and capacity to lead change at scale.

There is no one right answer to the multiple and enormous challenges of sustainable development. Instead, progress is won through multiple, multidirectional, collaborative efforts to evaluate, diagnose, design, and adjust in the pursuit of sustainability. The transition toward sustainability is under way. The world doesn't yet know all the right steps to take, but it's learning, and progress is being made ever more quickly. We hope you will find the ideas shared in this book to be helpful as you shape your own contributions to the sustainability transition.

Case Studies in Sustainability

This appendix expands on the short vignettes of chapter 1, presenting four very different cases of people working for changes that could promote the well-being of people, not just for today but for the long term. As we noted in chapter 1, these cases take place in different settings but are alike in that they provide examples of people trying to make good things happen, struggling to deal with unintended consequences, learning as they go, and continuing to push for progress. These case studies illustrate the complexity of the sustainability challenge, the importance of understanding complex social-environmental systems, and the role that human, social, technological, knowledge, and natural capital assets play in the long-term well-being of people.

LONDON: THE STRUGGLE FOR SUSTAINABLE DEVELOPMENT IN AN URBAN ENVIRONMENT

London today is widely recognized as a leading world city.[1] It regularly scores at or near the top of surveys about sustainability and quality of life in urban areas, racking up especially high scores for its international clout, its technological savvy, and its livability; doing well on economy, governance, and many dimensions of environment, but still struggling with air pollution and social cohesion.[2] The city is proud of its position, relatively self-critical of its shortcomings, and committed to doing better. The most recent plan for Greater London, an evolving document launched in the early years of the twenty-first century, is centered on a

> vision for the sustainable development of London . . . over the years
> to 2031 and beyond [in which] London should: excel among global

cities—expanding opportunities for all its people and enterprises, achieving the highest environmental standards and quality of life and leading the world in its approach to tackling the urban challenges of the 21st century, particularly that of climate change.[3]

That London can now aspire to such an ambitious vision for its future is remarkable—and instructive—when one considers the city's history. This consists of nearly two thousand years of apparently unsustainable developments, some very like those being experienced in today's most rapidly growing "new" cities. Over this period, the pursuit of short-term personal gains by Londoners repeatedly entailed long-term large-scale costs of social and environmental degradation that eventually became untenable. But people responded with a mix of political activism, scientific discoveries, technological inventions, social adjustments, and new forms of governance that—together with events occurring in the wider world beyond London—opened the way for the next round of development initiatives. These invariably led to their own surprises and readjustments, of which the current plan for Greater London is the latest installment.

How should we view the resulting whiplash trajectory of London's historical development from the perspective of sustainability? What should be learned from it that will help guide the ongoing global process of urbanization on a path toward sustainable development? These are the sorts of questions we hope this book will help you address. To provide some substance for our discussion of the London experience throughout the book, however, it will be useful to summarize a few revealing chapters in the city's two thousand years of development.

A Place by the Sea: London's Founding to the Norman Conquest

The Romans established the fortifications reflected in contemporary London's "square mile" around AD 50. The site they chose was, following the doctrines of Caesar's engineer Vitruvius, on a gentle slope well positioned for the sun to clear away mists and vapors from the lower grounds lying to the south.[4] Like most places that eventually become major cities, London was also founded on a river—a source of water, food, power, and transport. Its siting on the Thames River was special, however, placing London at the limits of where ocean tides affected the river's height and could thus also be used to help move goods *up*river from the open sea (about 80 kilometers to the west) as well as seaward with normal river flow. This "two-directional" river setting helped establish London as a major trading center. But it also meant that the city, along with much of its downstream hinterland, was vulnerable to flooding when strong North Sea storm surges coincided with exceptionally high tides. Efforts to build embankments and other protective works began in the earliest Roman times, continued through the medieval period, and included construction of the massive Thames Barrier gates in response to the flooding of the city in 1953. Efforts to retain the benefits of London's place by the sea, while reducing the related vulnerabilities, are playing a central role in the city's plans for dealing with climate change in the twenty-first century.

For its first thousand years of existence, London was repeatedly sacked, burned, occupied, and abandoned. By the time of the Norman Conquest in 1066, however, it had become a substantial city with enduring structures (e.g., the Tower of London), its own governing charter, and a population of 10,000 or more. The needs of that population were, at first, readily provided by the natural environment of the city and its immediate hinterlands. Water was abundant and clean. The city was surrounded by rich marshlands and agricultural clearings that not only provided food but were also a repository for human waste that was collected in the city and sold to the farmers as fertilizer. Beyond these intensively managed lands were extensive forests that furnished food, fuel, and timber for construction. Reflecting the central importance of these forest lands to social life, an elaborate though not particularly equitable system of traditions, rights, and laws emerged to govern who had access to them for what sorts of uses.[5]

Forest, Food, and Fuel to the Plague of 1348:
London in the Middle Ages

The early medieval period was a time of rapid population growth in England as a whole, as well as in London, which had become the country's principal city.[6] From 1100 to 1300, the population quadrupled, reaching a peak of perhaps 80,000 inhabitants (figure A.1). The chal-

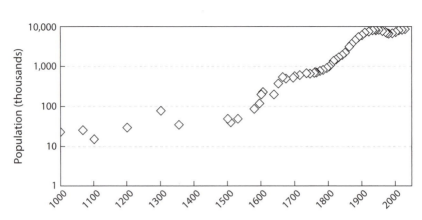

FIGURE A.1. London's population over the last millennium (thousands of people, plotted on a log scale). (1801–present is from Great Britain Historical GIS, University of Portsmouth, London GovOf through time | Population Statistics | Total Population, *A Vision of Britain through Time*, www.visionofbritain.org.uk/unit/10097836/cube/TOT_POP; 1700–1800 is from Landers, J. 1993. *Death and the Metropolis: Studies in the Demographic History of London, 1670–1830*. Cambridge, New York: Cambridge University Press; data for earlier period are from Keene, D. 2000. "London from the Post-Roman Period to 1300." In *The Cambridge Urban History of Britain*, edited by D. M. Palliser, 187–216. Cambridge: Cambridge University Press; Keene, D. 2012. "Medieval London and its Supply Hinterlands." *Regional Environmental Change* 12[2]: 263–81)

lenges of providing for the city's increasing material and energy needs, and of disposing of the resulting wastes, increased accordingly.

Some of the tightest early constraints on London's development involved water, both bringing it into the city and disposing of excess rainfall. To deal with the supply problem, numerous conduits were constructed to tap nearby ponds and lakes, as well as tributaries to the Thames, and to bring them all into the city center. Drainage was accomplished through unorganized runoff into the Thames and its tributaries, supplemented by the construction of additional "sewers" (modeled on those of Roman times) to capture runoff and convey it to the river. Industrial wastes, principally from breweries, tanneries, and the like, were dumped directly into the watercourses. Household waste, principally human excrement, was collected and disposed of in local cesspits (also called "cesspools" or "necessary chambers") that—as in earlier times—

were periodically emptied and the contents (fittingly called "night soil") transported to the countryside for use as fertilizer.

This waste disposal system worked, after a fashion. But it wasn't perfect. Because the flow of water down the streets or through the sewers to the Thames was irregular and could be interrupted, industrial wastes sometimes built up and rotted in the midst of the city. Pig sties were early banned from the streets because of their contribution to these disposal problems. Household cesspits were prohibited from connecting directly with the sewers. Overflows, however, were not uncommon. By 1189, regulations required that cesspits be located a minimum distance from property lines. Periodic complaints nonetheless occurred, as when the Carmelite Friars in 1290 felt obliged to petition that the "great stench" from the surrounding area made it impossible for them to carry out their religious duties. The problems became serious enough that by the mid-fourteenth century successive kings of England were moved to intervene. Edward III was particularly direct when he addressed the officials of London in 1357:

> Whereas now, when passing along the water of the Thames, we have beheld dung and other filth accumulated in diverse places in the said City upon the bank of the river aforesaid and also perceived the fumes and other abominable stenches arriving therefrom . . . [we] do command you that you cause as well the banks of the said river, as the streets and lanes of the same City, and the suburbs thereof, to be cleaned of dung and other filth without delay and the same when cleaned to be so kept.[7]

The commands, however, do not seem to have been backed up by the knowledge of how to accomplish the job, the resources to do it, or the governance structures to induce compliance. As a result, the stench would persist as an integral part of the London experience for another five centuries.

London's other great material need was for food. Some food was supplied by the Thames fisheries, but most was imported from farmers working a network extending 65–100 kilometers into the surrounding countryside. Increasing demand was met there through a combination of measures to increase yields on existing lands (e.g., through use of fertilizers, improved seeds, better management) and to expand the amount of land under cultivation. Together, these two supply responses managed to prevent serious hunger in much of Europe, including England and London, through the first half of the thirteenth century.[8] By the beginning of the fourteenth century, however, agricultural expansion had been pushed to the limit. Food shortages and malnutrition became increasingly chronic conditions. Moreover, the massive forest clearing needed to increase the land available for agriculture had also decreased the amount available for timber production. In the area surrounding London, there was little forest land left except that reserved by the aristocracy for recreation and game reserves. The resulting scarcity of accessible wood and other forest products had a number of impacts on social life. These ranged from suffering by peasants, who could no longer use the forest as their traditional subsistence backup when crops failed, to concern by royalty, who could no longer find the timber needed for their building projects. Perhaps more surprisingly, the growing scarcity of wood also brought on what may have been the first of London's many air pollution crises.

Since London's founding, relatively clean-burning wood had been its primary source of fuel for providing heat to homes and manufacturing (e.g., for smelting, metal working, and the production of lime for use in agriculture and building). But as forests made way for agricultural fields, wood and the charcoal made from it became increasingly scarce, distant, and expensive. At the same time "sea coal"—a low-grade, soft, high-sulfur fossil fuel initially sourced from coastal deposits—was becoming cheap and available. By the late thirteenth century widespread adoption of this new energy source in London was causing many complaints about "infected and corrupted air." In 1307 King Edward I intervened, prohibiting the use of sea coals in lime kilns because of the "intolerable smell" that resulted in "annoyance . . . and injury of their bodily health" for all in the vicinity. "Grievous ransoms" were authorized to punish offenders.[9] Complaints about fumes did subsequently decline, presumably reflecting improvement in air quality. In any case, air pollution did not again figure prominently in the documentation of London's affairs until well into the sixteenth century. It is tempting to attribute these apparent improvements in air quality to the effectiveness of informed government action. But Edward's royal decree coincided with the end of the medieval boom period. The declining demand for coal almost certainly helped achieve the king's goal.

Other problems, however, were on the rise. Despite the expansion of agricultural land that had so dramatically reduced forest cover, in the early decades of the fourteenth century, England—and indeed much of Europe—was struggling to feed itself. Food shortages, malnutrition, and even starvation became the norm. Population and economic growth slowed. Stagnation turned to collapse when the Black Death (almost certainly caused by the bubonic plague) reached London in 1348 after rapidly spreading through Europe from Asia.[10] Within eighteen months, an astonishing half of London's population was dead. The plague returned sporadically throughout the remainder of the century, reducing England's population overall by perhaps 40%. (The plague would remain a significant cause of death in London for the next 300 years.) The impacts of the Black Death on social well-being in the fourteenth century are incalculable. But just as the earlier increase in human numbers and activity had reshaped wetland and forest landscapes into agricultural ones, so the human depredations of the plague reversed that process. Untended embankments collapsed, letting the Thames reclaim much of its former broad floodplain (and in the process reduced the vulnerability of upstream London to storm surges).[11] Forests invaded the untended fields of London's hinterlands, soon providing enough wood to meet the much-reduced demand of the much-reduced London population. The use of sea coal—widely seen as less desirable than wood—declined accordingly. Complaints about air pollution declined correspondingly.[12]

Muddling through to the Great Fire of 1666: London of the Tudors and Stuarts

London did not recover its pre-plague population until the late fifteenth century. But by 1520 the city was again booming and toward the middle of the century had come to occupy a central position in the growing world trade network. By the end of the reign of Elizabeth I in the early seventeenth century, London had become the preeminent population center of England, with perhaps 220,000 inhabitants. Immigration rates were substantial, as more and more people from England and abroad came to the city to take advantage of its economic, political, and cultural attractions. Technical innovations abounded, including a water wheel that would pump water from the Thames to distribution points in the city for the next 240 years. The first private coaches also appeared, both speeding movement of people within the existing city and increasing the effective scale of the growing city. Governance innovations were also advanced. A Bill of Sewers, meant to provide for systematic cleaning and drainage of the city, was introduced in 1531.[13] Later efforts, reflecting awareness that the city was outgrowing its ability to function, sought to limit both the expansion of London's boundaries and the subdivision of its existing housing stock into ever-smaller, more densely packed units. These efforts almost certainly had some impact. But they were wholly inadequate to the rapidly growing challenges of a rapidly growing city.

For example, as booming London once again pressed its surrounding forests to supply land for agriculture and timber for construction, wood again became too scarce or expensive for use as a heating fuel. Sea coal again became the dominant fuel source, once more leading to complaints about foul emissions. This time, the king called on the newly emerging science establishment to assess the causes and consequences of troubling smokes. Scientists got parts of the air pollution story about right by today's standards, concluding that smoke from coal-burning industries caused material corrosion and damage to the lungs of city dwellers. And they advanced logical proposals for mitigating damage: precombustion treatment of coal to limit smoke emissions, and movement of industries to the periphery of London. But there was no acute crisis to focus attention, and no mechanism in place for the slow accumulation of knowledge to produce action. London got smokier.[14]

The inadequacy of the system for solid-waste disposal once again became apparent. The king wrote to London's mayor:

> The king hath noticed that the ways in and about the City and liberties were very noisome and troublesome for passing, in consequences of breaches of the pavements and excessive quantities of filth lying in the streets. They require him, by the king's express command, to take effectual steps for the complete repair of the pavements and the removal of all filth.[15]

In short, the stinks of old were back. The king, as before, could command them to go away but was ultimately disappointed with the result of his orders.

The combination of increasingly crowded living conditions; the buildup of human excrement in backyards, alleys, and water supplies; and the growing trade with the rest of the world set the stage for an upsurge in infectious disease epidemics.[16] These included not only the plague, which recurred at frequent intervals, but also an often undifferentiated and increasingly deadly mix of influenza, scarlet fever, tuberculosis, typhus, typhoid, diphtheria, measles, whooping cough and—above all—smallpox. As a result, it was not uncommon for London to lose 10% or more of its total population to epidemic disease outbreaks in any given decade.[17] Then, in 1665, the Great Plague struck with a violence not seen since the first plague of the fourteenth century. With vastly more people, rats, and fleas available in London to support the epidemic, the human toll was unprecedented: as many as 100,000 Londoners may have died over the seven months of the outbreak, with more than 7000 per week perishing during its peak. Society at the time had no reliable understanding of the plague's causes and therefore was left flailing about with a variety of largely ineffectual measures to combat it. These ranged from mass killings of dogs and cats (which must have caused the fleas that actually transmitted the disease to look for new hosts), to packing up and fleeing the city (which hastened the spread of the disease), to a breakdown of social networks as people refused to help sick servants, friends, or family. Samuel Pepys glumly noted in his diary: "the plague [is] making us cruel as dogs to one another."[18]

Whether London's response to the Great Plague would have differed substantially from its response to previous epidemics will never be known. For the Great Plague was followed in the next year by the Great Fire: a conflagration fueled by the city's tightly packed and almost wholly wooden structures. London had experienced fires before, but that of 1666 was the worst, destroying a third of London and more than 80% of the core City, where finance and trade were concentrated. Surprisingly few people died. But perhaps a quarter of the population that had survived the Great Plague of the year before was left homeless. Many people spent the following winter camped in tents in fields surrounding the city. Others fled the double devastation of plague and fire, never to return. London's prospects seemed bleak. An increasingly depressed Pepys wrote: 'the City less and less likely to be built again, every body settling elsewhere, and nobody encouraged to trade.'[19]

A Waste of Life: London's Battle with Disease in the Eighteenth Century

London struggled through the remainder of the seventeenth and the first half of the eighteenth century with a slow, piecemeal reconstruction. Various grand plans for a new city came to nothing. The London that rose from the ashes had much the same street plan as the old one. That London was, however, less flammable, as stone replaced wood as the preferred building material. In the city center, financed by a new coal tax, thoroughfares were paved and widened, drainage sewers were added, and several grand structures were built that still stand today, including Guildhall and numerous churches by Christopher Wren, including St. Paul's Cathedral. Reflecting on the result, one giddy resident heralded London as "not only the finest, but the most healthy city in the world."[20]

Beautiful it may have been, at least in the newly built center city if not the extensive slums. But healthy it was not. In fact, London experienced a general *rise* in mortality throughout the seventeenth century and had, by the second quarter of the eighteenth century, become an unprecedentedly unhealthy place.[21] More than a third of the babies born in the city died in infancy. Overall life expectancy was a mere eighteen years, barely half of what it was then in England as a whole, and less than a quarter of what it is today (figure A.2).

One historian summarized the consequences as follows:

[F]or most of the eighteenth century London is dominated by a sense of the "waste of life."[22]

The causes of London's deteriorating health in the early eighteenth century were unknown at the time and remain difficult to disentangle today. It seems fairly certain, however, that no major new diseases arrived in London during this period. Neither long-term trends in climate nor food shortages seem to have played a central role. More likely, the major contributor to the decline of health in London during this period

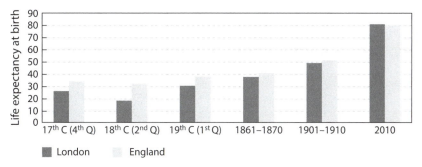

FIGURE A.2. Life expectancy at birth in London and England (years). (Pre-1850 London data are from Landers, J. 1993. *Death and the Metropolis: Studies in the Demographic History of London, 1670–1830*. Cambridge, New York: Cambridge University Press, fig. 5.3 and table 4.10. Contemporary London data are from Public Health England database: www.lho.org.uk, especially Baker, A., G. Findlay, P. Murage, G. Pettitt, R. Leeser, P. Goldblatt, J. Fitzpatrick, and B. Jacobson. 2011. *Fair London, Healthy Londoners?* London Health Commission, London; and Greater London Authority Demography Team. 2010. *Infant mortality: 2002–2008* No. Update 09–2010. London: Greater London Authority. Pre-1850 England data are from Wrigley, E. A., and R. Schofield. 1989. *The Population History of England, 1541–1871: A Reconstruction*. Cambridge, New York: Cambridge University Press [tables 7.15: life expectancy at birth, by quartiles, and 7.19: infant mortality, by half century]. Data for London and England 1850–1910 are from Woods, R. 2000. *The Demography of Victorian England and Wales*. Cambridge; New York: Cambridge University Press)

was the growing density of people with little immunity to the many diseases that were endemic to the city. This trend had two, and possibly three, interacting components:

- First was the extraordinary flow of young immigrants from the countryside. Perhaps three quarters of Londoners at the time had been born somewhere else, primarily in the rural areas of England, Scotland, and Ireland.[23] Most of these immigrants had experienced during their rural childhood relatively few encounters with the microbes they would meet in London and thus had neither been selected for, nor developed, immunity.
- Second was the dismal condition of London's housing stock. The stagnant economy had slowed new house construction to a trickle. As a result, existing buildings were subject to decay and "internal coloniza-

tion": repeated subdivision into smaller and smaller rooms packed with more and more people, rats, and lice.

- Third may well have been depressed rates of breastfeeding. The extent of decreased breastfeeding in this period is unclear but seems to have been substantial. Causes may have involved changes in demands of work, general illness, worsening nutrition, or social customs. The consequences, however, were stark. Breastfeeding provided infants who received it with food that was nutritious, free from contamination, and laden with antibodies and other antimicrobial substances passed on from the mother. Children lacking these benefits due to limited or no breastfeeding may have been 5–10 times more likely to die as infants than those exclusively raised on breast milk.[24]

Together, these three components—a flow of immunologically naïve immigrants, increased crowding in dilapidated housing, and a lack of breastfeeding—created in London a perfect environment for efficient transmission of the many diseases endemic in the city. The "waste of life" that characterized the city in the eighteenth century was the result (figure A.3).

But if London of the early eighteenth century was a death trap, it was also the center of a generally prosperous England's government, its greatest port, and its largest manufacturing center. Though it had deep and persistent problems, London was also growing a vibrant civic life. Coffee houses, newspapers, and political debate proliferated. Business prospered. And social reformers began to address what they saw as the problems of the poor, providing workhouses to alleviate poverty, and hospitals to support newborns and treat the sick, and generally striving to make the city a better place to live.

One of their most effective efforts focused on smallpox. In terms of cumulative deaths, this was almost certainly London's single most lethal disease of the time. A fairly effective preventive treatment involving subcutaneous grafting of pustules or scabs from an infected individual had been known for centuries in other parts of the world with a longer history of smallpox. But how these "inoculation" or "variolation" treatments worked was not understood. They were therefore rejected as without merit by the scholars of the Royal Society and London's medical profession. Eventual adoption in London owed less to

FIGURE A.3. Death, Dirty Father Thames, and King Cholera. (Photos © Punch Limited)

science than to the determined advocacy of Lady Mary Montagu. As wife of England's Ambassador to the Sublime Porte, Lady Montagu had seen the effectiveness of variolation on her journeys. When her testimony did not change minds of establishment science, she had the treatment successfully applied to her own daughter in the presence of physicians of the royal court. Faced with this evidence, and a high demand for protection from smallpox, the medical establishment undertook in the early 1720s some systematic experimentation (on prisoners and orphans). The generally positive results of these questionable experiments finally led them to endorse the variolation procedure. The slow adoption of variolation in both London and the hinterlands from which its immigrants came helped, by 1760, turn the tide against smallpox and nudge along a slow decline in deaths attributed to it.[25] Progress accelerated once true vaccination was added to the arsenal of anti-smallpox measures owing to the pioneering efforts of Edward Jenner in the early nineteenth century.[26] The Royal Society rejected his initial paper on the subject, but he and a growing number of others persuaded by the evidence persisted.

By the middle of that century, the cumulative efforts of Montagu, Jenner, and other pragmatists—with only a little help from scientific theory—had relegated smallpox to a relatively minor role as a cause of death in London.[27] Coupled with the departure of the plague (which had largely disappeared from London in the wake of the Great Fire), with the better nutrition and housing that came with a recovering economy, and with continuing efforts of civic reformers and medical practitioners, the health of the city finally began to improve. A Londoner born in the 1820s had twice the chance of surviving infancy, and almost twice the life expectancy of one born a century earlier.[28] With its demographic transition firmly under way, the city embarked on a century of rapid population growth. This would take it from about one million inhabitants in 1800 to more than six million people (and 300,000 horses) by 1900.

Water, Waste, and Sewers up to the Great Stink of 1858: Victorian London

As a rapidly growing London took an increasingly dominant place on the world stage, the city's physical infrastructure and domestic governance processes lagged behind. London's chronic shortage of quality housing has already been noted, and it certainly worsened over this period. But nowhere was the stress of rapid urban growth clearer than in London's losing battles with its enduring challenges of managing its water and solid wastes. With the exception of the very wealthy, the million inhabitants of early nineteenth-century London still relied for disposal of human excrement on essentially the same system of cesspits, night-soil men, hand pumps, and open sewers that had been developed to serve a medieval city one-fortieth the size.[29] For water supply, poorer districts continued to rely on the shallow wells that had served the city for centuries. Now, however, the growing demand for water was met by private companies intent on generating returns for their investors. They did so by encouraging higher and higher water consumption. But they could find no profit in handling the orderly disposal of the waters they provided, so sewage systems remained undeveloped.

The farms that once received most of the contents of the cesspits as manure were increasingly far removed from the city, making disposal of wastes there more costly and less likely to occur. (That recycling system would collapse almost entirely after 1847, when an expanding world trade system began to capitalize on Alexander von Humboldt's earlier discovery that huge Peruvian deposits of bird droppings, or guano, could be used as very effective agricultural fertilizer. Mined by indentured Chinese labor and shipped to England, guano killed the demand for London's night soil by farmers.)[30] With no one motivated to remove the wastes, London's system was overwhelmed. Its cesspits overflowed into the other water management channels, commingling wastes with surface drainage water in the sewers, mixing it with many of the city's water sources, and emitting an ever-more-pervasive stench.

Several responses to this deteriorating situation were possible. Unfortunately the responses initially adopted were guided by the prevailing science of the day, which was wrong. Confusing correlation with causation, scientists argued that disease was associated with airborne "miasmas" arising from rotting fecal material and other wastes. This theory led to, or reinforced, a practical engineering emphasis on protecting people from tangible stenches rather than protecting water supplies from intangible contamination. In particular, it encouraged widespread adoption in the early nineteenth century of the new "water closet" technology that swept foul-smelling human household waste away to "somewhere else." The theory supported the lifting (in 1803) of a long-standing prohibition on connecting household drains directly to drainage sewers and thence to the "somewhere else" that was the Thames. It was eventually used to justify the requirement (in 1848) that such connections be made for all households to reduce dangerous back-alley odors. The resulting success in flushing London's stenches away from houses meant that the Thames rapidly became a common cesspool for all of London, essentially replacing many of the thousands of backyard cesspools that had previously held the city's wastes (figure A.3). The miasma science of the day provided no reason to worry over the fact that this new Thames cesspool had somehow lost all its once-plentiful fish life but still served as the source of much of the city's drinking water.

Better knowledge, based on the germ theory of disease, would eventually show why dumping fecal matter into one's water supply had not just unaesthetic consequences but also unhealthy ones. But general acceptance of this healthier perspective on how to organize human activities would need to await the discoveries of Pasteur, Koch, and their colleagues, beginning in the 1860s. In the meantime, pressures to clean London of its centuries-old stenches persisted, driven by concerns over the continued presence of many old diseases in London and the arrival in England in 1831 of a deadly new one from Asia: cholera (figure A.3). Cholera was a terrifying disease, even for a city accustomed to epidemics. It would kill by dinner people who had seemed healthy at breakfast, and struck down rich and poor alike. London's first epidemic surged through the city in 1832, killing more than 6000. Medical science was initially unable to offer alternatives to miasmic accounts of the disease's origins or propagation. Remedial treatments thus explored all sorts of ad hoc and ultimately ineffectual measures. And public health efforts continued to focus on separating people from the stench, rather than people from their feces.

Fortunately, some measures undertaken to improve the smell and other aesthetic qualities of London's disgusting drinking water coincidentally also reduced the risk of waterborne disease, including cholera. For example, some of the private water-supply companies competing for Londoners' business developed sources from relatively sweet-smelling (but also disease-free) tributaries and springs. Others introduced the innovation of slow percolation through sand for the purpose of removing waste before it could stink. But this filtering process accidentally and unknowingly also removed most bacteria before they could kill, which is why it remains widely used today. Such measures doubtless contributed to the reduction of both stench and disease in the early nineteenth century. But they were not widely enough adopted to spare London from a return of cholera in 1848–49 and 1853–54, with each outbreak killing more than 10,000 people.

By the time of these midcentury outbreaks, however, what today would be recognized as modern science of statistical inference had begun to take form. Drawing from the extensive records of the locations and causes of deaths kept by the city, scientists were able to make empirical headway on the disease front, even under the burden of incorrect miasma theories regarding the origins and propagation of those diseases. Most famously, the physician John Snow concluded from data he mobilized on deaths occurring during the 1848–49 cholera outbreak that the disease was more likely to result from ingestion of a particle than from inhalation of miasmic air. Snow's statistical arguments did not immediately carry the day in scientific circles. But along with politically forceful arguments from the royal palace that the Thames water stank and tasted terrible, plus the persistent public fear of another cholera epidemic, his findings almost certainly helped gain passage of the Metropolis Water Act of 1852. The act required that by 1855 all of London's drinking water taken from the Thames would have to be sourced from above the limits of its tidal reach (and so upstream of the city's pollution) and would have to be passed through the slow sand filters that up to then were only in patchy use.

The act was too little, too late to spare London another cholera epidemic in 1853–54. But intense study of the mortality patterns exhibited in that epidemic by Snow and the scientists on the London Health Board finally showed beyond reasonable scientific doubt that people who drank putrefied water during cholera outbreaks were much more likely to die than those who drank relatively pure water, even when they lived in common neighborhoods exposed to the same miasmas of air and other environmental conditions. In a scientifically admirable, if practically exasperating, indication that they correctly understood the limits of epidemiological statistics, however, the scientists of the health board were careful to stress that they had not shown that bad water *caused* cholera and had no theory that would explain such causation.

They did, however, have compelling evidence of the greatest practical significance: substituting clean water for filthy water radically reduced the chances of contracting cholera and other diseases. They also presciently pointed out that while implementing the Water Act's requirement to use only the upper Thames as a source might help limit the immediate risk of cholera, it "remains but imperfect and precarious while . . . [upstream] populations exercise a right of sewerage into the drinking water of London."[31] Private water suppliers exploited the scientists' caution to press their claim that Thames water, though "nourishing a population of animalcules, would not be noxious to health."[32] Politicians dithered about the administrative and financial hurdles involved in establishing clean water supplies. John Snow famously demonstrated that removing the handle of the contaminated Broad Street pump could arrest the local spread of the 1853–54 cholera epidemic. But once that epidemic had passed, little action followed, despite what is now understood to be broadly correct and potentially useful scientific findings provided by the health board.

Action, when it did come, was precipitated not by scientific knowledge but by cataclysmic events. In this case, a particularly hot June in 1858 transformed London's communal cesspool—the river Thames—from its chronic state of mere foulness into what would become notorious as the Great Stink. A contemporary wrote:

Such was the overpowering smell from the Thames, that the curtains of the Commons were soaked in chloride of lime in a vain attempt to protect the sensitivities [of the rich and powerful].[33]

The future prime minister of England, Benjamin Disraeli, was seen fleeing the chamber, decrying the "Stygian pool" that was sloshing back and forth with the tides through the middle of what was supposed to be the greatest city on earth. When cooler weather allowed it to reconvene, Parliament acted with unprecedented speed. By summer's end it had authorized a massive and innovative program of sewer construction. Within seven years this huge infrastructure project would capture much of the household sewage of London and divert it to discharges downstream of the city. The reasons cited in the parliamentary debates reflected not merely the unaesthetic qualities of the stink (which they had lived with, at one intensity or another, for all their life) but also what was now a mix of beliefs that by eliminating the miasma of foul air, or the putrid water, or both they might also reduce the incidence of cholera and other diseases that threatened to kill them.

Subsequent events seemed to justify their hopes. Cholera returned to London for what proved to be the last time in 1866. But it claimed only half the victims of previous outbreaks, with deaths concentrated in the parts of the city not yet fully served by the new sewer system. Mortality in London from typhoid fever and diarrhea declined as well, even as cities without sewer systems experienced recurrent epidemics. Victorian England, already smug for many reasons, celebrated its new sewer system as "the most extensive and wonderful work of modern times."[34]

A major and daring accomplishment it was, and one that most certainly improved the quality of life in London. Much of its basic structure remains in place today. The engineering marvel that was London's sewers did not, however, solve the problem of water and waste disposal, but just pushed them away—downstream and into the future. (Later measures would push them farther still, first barging the wastes collected at sewer outfalls to the ocean for dumping, today discharging them into the global atmosphere through incineration.) London's sewer innovations of the 1860s didn't solve its disease problem, either. (With the robust germ theory as it is understood today, it is clear that the big assault on waterborne diseases was already—if not altogether consciously—well under way with the passage of the 1852 Water Act and its provisions for clean water sourcing and filtering.) In fact, the most important innovations surrounding the Great Stink and its resolution may not have involved how to engineer health but, rather, how to govern it.

Before the Great Stink, responsibility for the water supply and waste systems was distributed among hundreds of local authorities, each with its own locally attuned rules and technologies. This self-organized patchwork may have been adequate, even superior, when London was smaller. But as the city grew, what had been local strength became a regional weakness of turf fights and incompatible piping. Crucially, however, the various acts surrounding the Stink consolidated the local authorities under a new Metropolitan Board of Works. The board was given the authority and resources to impose solutions at scale that previous measures had lacked. Above all, in the wake of the Great Stink a system was developed that allowed the board and various municipalities of the greater London region to raise revenue for their waterworks through borrowing backed by property taxes and water-use levies. These governance innovations provided the foundation on which subsequent public works measures to manage social-environmental interactions were built over the coming decades.[35] Together with continuing advances in the science and engineering of water supply and waste removal, they bequeathed modern London both better human health and a better Thames environment than anyone in the nineteenth century had ever imagined, with productive fisheries, occasional swim races, and rarely a stench to its name.

Persistent Air Pollution and the Great Smog: London's Twentieth Century

Preoccupied with their challenges involving water, cholera, and fecal matter, Londoners of the nineteenth and early twentieth centuries seem to have taken their lack of air quality for granted, referring to their city, not entirely unaffectionately, as The Smoke. (This is sadly ironic, for we know today that the health impacts of urban air pollution—particularly the particulate matter that constituted the London smokes—are substantial.)[36] Sir Arthur Conan Doyle, for example, seems to have been describing rather than criticizing London when he dispatched Sherlock

Holmes off on chases through the "dense yellow fog" that he saw as characteristic of the city in the late nineteenth century. Those traveling to London from the country were less sanguine. So were social critics such as Charles Dickens. His Esther of *Bleak House*, for example, asked on her arrival in London several decades before Holmes roamed its streets "whether there was a great fire anywhere? For the streets were so full of dense brown smoke that scarcely anything was to be seen."

As noted earlier, fumes and smoke had always been part of London life (and death). The complaints of the thirteenth century were largely due to industrial use of coal, particularly in the production of lime. As coal had been increasingly adopted not only by industry but as a domestic heating and cooking fuel in the late sixteenth century, Londoners had thrown themselves into the normal work of pollution control: making the bad stuff go somewhere else. In this case, the "somewhere else" was the air of London, as indoor air pollution was shifted outdoors through increasing use of low household chimneys.[37] Complaints began to build again, scientific studies were conducted, but not much changed. A century later, the invention of the steam engine resulted in a rapid increase in coal use throughout the city. Some regulations to limit emissions were put in place, but these were largely focused on a relatively few industrial installations. The idea that Londoners themselves, devoted as they were to their individual hearths and chimneys, were a big part of the overall air pollution problem got little traction. As a result, the same combination of a surge of London's population and an industrial revolution that had overwhelmed the city's ability to provide acceptably healthy water to its people in the nineteenth century overtook its ability to provide acceptably healthy air in the twentieth century.

A series of increasingly severe "pea soup" fogs in the first decades of the twentieth century brought about increasing concerns for the impact of deteriorating air quality on the city, its commerce, and its inhabitants' health. The Second World War, however, turned people's attentions to more immediate concerns. The war affected London in multiple ways, as it did the rest of England and the world. Most immediate for the city, however, was the Blitz: the strategic bombing of the city by Germany. At least 20,000 Londoners were killed, a second "great fire" ravaged the city, and more than 300,000 of its houses, together with much of its industrial base, were destroyed. Many people fled the city, and the population continued to fall over the next half century.

But London once again showed itself able to recover and rebuild in the wake of disaster, drawing on people and other resources from a nation and foreign friends for whom London had become simply indispensable. This rebuilding began to grapple seriously with housing and green-space issues, and once again air pollution returned as a subject of discussion. Then, in December 1952, London experienced a week-long "killer smog" that was at least as severe as anything in today's headlines about rapidly growing cities of the developing world. Contemporary accounts refer to

> the worst fog that I'd ever encountered. It had a ... strong, strong smell ... of sulfur. ... Even in daylight, it was a ghastly yellow color. ... You literally could not see your hand in front of your face.[38]

Transportation stopped. A performance of *La Traviata* was canceled because of poor visibility *inside* the opera hall. The black smoke permeated everything, from the stacks of the British Library to the underwear of householders.[39] This single pollution episode came to be known as the Great Smog and may have killed as many as 12,000 Londoners[40]—more than half the number killed by the entire wartime bombing blitz and comparable in numbers (though not proportions) to the toll of individual plague and cholera outbreaks of earlier times (figure A.4).

Unfortunately, the magnitude of the death toll was not fully appreciated for another 50 years, when careful scholarship disentangled how this acute episode of air pollution increased the death rates from chronic respiratory conditions. Action therefore did not follow the Great Smog of 1952 as quickly or forcefully as it had the Great Stink of 1858. But in its wake, air pollution would never again be accepted as an inevitable condition of living in London, and regulations to reduce it finally were accorded broad public support. Parliament eventually passed a Clean Air Act in 1956. This included for the first time efforts to control domestic as well as industrial production of smoke. It established "smokeless" zones, thus inducing fuel switches away from the dirtiest coals and effectively reducing particulate emissions. It did not directly address sulfur or other emissions, though some of these declined as well as a result of the smoke-driven fuel switches. Systematic

FIGURE A.4. Piccadilly Circus in the Great Smog. (Photo © Hulton Archive/Getty Images)

progress on dealing with these other pollutants did not occur until the last decades of the twentieth century. Some of that progress was simply due to the continuing clearance and replacement of London's old housing and to the export of much of its manufacturing industry to other parts of the world. But efforts to improve London's air quality, like similar efforts elsewhere, benefited greatly from novel synergies that began to take shape in the 1980s between programs of transnational environmental research and programs of transnational environmental governance. For the first time, those programs systematically brought science to bear on the shortcomings of development strategies that dealt with wastes by pushing them elsewhere. Work on acid rain (involving many of the same pollutants as a the Great Smog) led the way, followed quickly by international programs integrating monitoring, research, and governance to address the challenges of stratospheric ozone depletion, climate change, and a still-expanding array of other pollutants.[41] These global scientific programs combined with regional initiatives by the European Union began to pull Britain toward more aggressive environmental policies. They simultaneously provided leverage for local London reformers, who succeeded in pushing forward homegrown innovations such as the inner-city congestion charge on automobile traffic.

Whether such coalitions of global science and local action in the environmental sphere can be expanded, or learned from, to provide comparable support for understanding and addressing the social dimensions of sustainability remains to be seen. What is clear is that today's urban centers are the crucibles in which some of the most integrated pursuits of sustainability are taking place. One of the most inventive such crucibles is the city of London.

FARMER-MANAGED IRRIGATION
SYSTEMS IN NEPAL

Growing food in Nepal is a challenging prospect.[42] The high altitude and mountainous topography in much of the country precludes many crops altogether, and the relatively small area of land conducive to agriculture is already under cultivation.[43] Growing more food in Nepal, then, means improving crop yields (the amount of food produced per unit area of land used to grow it). One key way to increase yields is by improving the use and reliability of irrigation systems.

Irrigation systems are especially important in Nepal because of the characteristics of its climate.[44] Summer monsoon rains fall heavily from June through September each year, caused by winds that carry moisture from the warm Indian Ocean inland to landlocked Nepal and release it as they encounter high mountains. The other eight months of the year are very dry. It is therefore essential for Nepali farmers to make efficient use of the water while it is available and to capture what they can for later use. In the monsoon season, floods and landslides are common, and farmers have to work hard to maintain the condition of their fields and the irrigation systems that serve them.

Initial Failures: How Technological Improvements
Were Stymied by Social Realities

Since the 1960s, government and non-governmental agencies have tried to build and improve the performance of irrigation systems in Nepal.[45] In the 1990s, a group of American and Nepali policy analysts evaluated some 150 irrigation systems in Nepal, classifying them by their levels of technological improvements. They noted in particular whether each system had permanent headworks (the structure at the head, or beginning, of a particular irrigation system), canals lined with cement or stone, or both of these. They similarly classified each irrigation system in terms of its physical condition (how much maintenance it needed), the amounts of water delivered to those within it, and the productivity of crops grown within it. What they found was a shocking result: the productivity of agriculture was actually highest in those systems with absolutely no technological improvements. Lining canals did help deliver more water to farmers; the physical condition of these systems tended to be better than those not lined. But the worst-performing irrigation systems in all areas were those with the permanent engineered headworks.

How could these respected engineering projects not deliver improvements to farmers in Nepal? The answer lay in key elements of farmers' social systems that the agencies had overlooked. Because the sophisticated equipment used to create many of the headworks was difficult to maintain and operate, the government agencies took over authority for the new irrigation systems. Farmers themselves were not required to maintain the systems. With no need for collaboration, cooperation, or shared labor by up- and downstream farmers, farmers close to the headworks stopped worrying about maintaining good relationships with downstream farmers and started withdrawing more water and leaving less for their downstream neighbors' use. As a result, agricultural productivity of the system as a whole declined.

Overcoming Challenges through
Community Adaptation and Cooperation:
The Story of a Village in Kavre

Many communities in Nepal have not had any outside assistance at all and have struggled to create irrigation systems that can provide a more reliable supply of water for their crops. There are many remarkable stories of how small-scale subsistence farmers have overcome very adverse conditions—including extreme poverty, inaccessible geography, and complicated hydrology—to build their own farmer-managed systems. What follows is one such story.

In the valley to the northeast of Kathmandu, in the Kavre district, there is a small village of about two hundred inhabitants. The people who call this home settled the land in the 1950s, shortly after the government carried out a land reform that provided community titles to settlers interested in farming the land in the area. In 1970, almost all the families in the village still grew most of own their food, mainly maize and millet. Their farming techniques were very simple: they relied on Zebu oxen to plow the fields,[46] had no irrigation, and applied no chemicals or fertilizers other than animal waste on their crop fields. At the time, the average family produced enough food to last for about five months each year. To make ends meet, the men of the families would go to the cities to look for work during nonfarming season and would send money back to the village. In the 1970s, the average family had two adults and five children, but only about half of all children in rural Nepal were able to finish primary school.[47] One of every four children died before reaching the age of five. Life expectancy for rural Nepalese was 39 years.[48] Yet, in the early 1980s things in this village took a turn for the worse.

In 1983, harvests started to decline. The reason seems to have been a combination of erratic rains, increased soil erosion, and mudslides. A low point occurred in 1986, when villagers produced enough food to last only two to four months. That year women and children took to the forest to search for wild yam, which they would need to boil all night to make it edible by morning. Somehow no one died from starvation, although no one had enough to eat. Most families were just hanging on, barely surviving.

The villagers in Kavre had been unsuccessful in getting the Nepali government to help them build an irrigation system. Their standing request with the district government had languished for twenty years, with no help arriving. According to one of the village elders, the reason for the district council's unresponsiveness was that the village did not have a "man at the top" to help them.

The villagers knew that they could not continue like this; their leaders decided they needed to build that sought-after irrigation system, with or without the district government's help. A source of water, from a small stream, was within 2 kilometers of the village. If water from the stream was brought by canal to the village fields, it would provide enough water to double the number of harvests and allow farmers to diversify their crops to provide more security and, it was hoped, better income as well. Everyone agreed that working together to build an irrigation canal would be far better than the status quo and a more attractive option than abandoning their village to move to the lowlands.

But how could they design, build, and maintain a new irrigation system? And how would they do this without government support? What gave them confidence to succeed?

First, the villagers knew their village lands. They knew the place, its people, and its natural resources—including the characteristics of the stream—better than anybody else. The knowledge of how the stream behaves and how it changes with the seasons would be very important in designing and constructing the irrigation canals. And perhaps most important, they knew each other well, shared a common history, and trusted each other's commitment to fight for the survival of their village. These assets notwithstanding, their construction project faced three major challenges.

The first challenge was to get the necessary equipment and materials to do the work. They needed tools and also cement for certain portions of the headworks of the canal. They applied for credit to a new agricultural development bank, which gave them a small loan for the tools and materials. The farmers had to agree to provide all the labor and to repay the loan within five years.

The second challenge was of a more technical nature. To get the water from the stream to the fields, the planned canal line would have to cut through large rock outcroppings. Without sophisticated tools and machinery, how would they get through the rock? Rather than delaying the project by asking for help from outsiders, they did it by hand. Suspended by ropes around their waist, and equipped with only hammers and chisels, young men of the village chipped away at the huge rocks at a rate of about 1 foot per day; several weeks later, the canal line was through. This feat is, to this day, celebrated as one of the great achievements of the village.

Finally, the third challenge was to maintain the high level of cooperation that was required to pull off this project. The task required all members of the community to pitch in and contribute to the collective

endeavor. They did this by volunteering their time and labor, by negotiating and interacting with the bank, by cooking for the workers, or by
looking after small children. All collaborative efforts face similar challenges, and the larger and more diverse the group, the more difficult it
is to motivate individual group members to contribute to the collective
efforts. The high level of trust and community cohesion in this village
helped reduce the severity of this problem, but the temptation to let
others do most of the work—and "free-ride" on their efforts—is always
there. The villagers dealt with this issue by setting down clear rules
about what was considered a fair contribution by each household to the
village project. The village leaders then devised a system for monitoring and enforcing these rules. The implementation of these arrangements furthered the cooperation and collaboration needed not just to
build the system but to maintain it.

After about a year of the villagers' working together, the irrigation
system was completed and brought much-needed water to 20 hectares
of farmland. Farmers managed to increase their food production to
provide sufficient food for eight months. They paid off the loan in time
and have operated and maintained the canal by themselves ever since.

The Government Tries Again: Improving Irrigation
with Farmer Participation

Not all communities are lucky enough to have the leadership and cooperative potential that the village in Kavre enjoyed. But as national and
international organizations continued their efforts to improve irrigation systems in Nepal, they began to learn from the cooperative successes of villages like the one in Kavre. One project, in particular, is
instructive in terms of learning from the successes and challenges of a
participatory effort.[49]

In the late 1980s, the Water and Energy Commission of Nepal part-
nered with the International Irrigation Management Institute to develop a program for 19 irrigation systems in the Indravati River watershed near Kathmandu.[50] Drawing on scientific analyses and evaluations
of earlier failures, this program introduced several innovations. Farmers got to choose whether they wanted to be involved. If farmers wanted
funding, they had to agree to provide some of the labor and materials.
Participating farmers received farmer-to-farmer training from some of

FIGURE A.5. Farmer-constructed and farmer-managed irrigation systems. (Photo © Helvetas Swiss Intercooperation; Local Infrastructure for Livelihood Improvement: nepal.helvetas.org/en/our_projects/lili.cfm)

the most productive irrigation systems in Nepal. And each farming group was required to write a set of "working rules" that would govern decision making in their irrigation system (figure A.5).

What were the long-term impacts of this project? At the beginning of the project, an international team of researchers led by Wai Fung Lam and Nobel Prize winner Elinor Ostrom collected information on how each system was working (for example, the size of the irrigated area, technical efficiency of the irrigation infrastructure, water access and availability, and cropping intensity levels) and also interviewed farmers to understand their perspectives. They repeated their surveys in 1985, 1991, 1999, and 2001.

The team found mixed results. The project succeeded in increasing the size of irrigated area in the majority of the systems. However, nearly all the systems experienced deterioration in the technical efficiency of the system; that is, water was not being delivered as effectively as it could have been if all parts were working as designed. Nevertheless, most of the systems continued to make water available to farmers when they needed it. In terms of crop productivity, a more complicated picture emerged; in some systems, crop productivity improved over time, while in others it declined or did not change. Finally, thanks to the project, no farmers experienced water deprivation. These results indicated that the project was successful in improving farmers' interactions with one another, which led to more equitable water allocation and eliminated water deprivation; however, the infrastructure steadily deteriorated, and crop output was uneven, suggesting that a complex mix of factors was affecting farmers' use of the irrigation systems.

In trying to understand these results, the researchers wondered about the factors that affected differing outcomes among the systems. They tested the relative importance of several factors: government funding for infrastructure improvements, the existence of written rules, imposition of fines, leadership, and collective action among the farmers. Their results suggested that both collective action and written rules are necessary conditions for long-term improvement in crop productivity in the project area. In contrast with results from its own previous research,[51] the team found that imposing fines for rule breaking was not necessary in the case of this project, perhaps because of the strong social capital fostered by its initial setup. Continued funding from the national government was also not necessary for the irrigation systems to work well.

The researchers' findings indicate more generally that giving farmers a voice in the design of irrigation projects is essential for positive results. Maintaining irrigation infrastructure is important, but equally important is "human artisanship," or the ability to work together for mutual improvement (figure A.6).

Learning from Complex Systems:
Lessons from Nepal

According to some experts, there may be as many as 100,000 farmer-managed irrigation systems in Nepal, each with its own story of how farmers cooperated to develop a common resource that would benefit all. These systems experience a range of problems, including degradation of infrastructure, outmigration of farmers to cities, internal con-

FIGURE A.6. Farmers and their irrigation project. (Photo © FMIST Nepal; Farmed Irrigation Systems Promotion Trust, fmistnepal.wordpress.com)

flicts, pollution and silt deposition due to upstream land use and other activities, and lower water quantity due to warmer and drier weather in recent years. Scientific studies from the social, natural, and physical sciences have helped elucidate the multiple factors that have conspired to produce these problems, and also illustrate reasonable responses. While these systems are farmer-managed systems, scientists, governments, and NGOs have a role to play in helping them evolve to meet current and future needs. Both farmers and scientists are the keepers of another important resource: the lessons learned from past interventions in particular places and times.

THE YAQUI VALLEY: MOVING TOWARD SUSTAINABILITY WITH IMPERFECT BUT PERSISTENT INTERDISCIPLINARY RESEARCH

This case study tells a story of a research and action effort that engaged a team of scientists and decision makers over about fifteen years.[52] The project was initiated and led by one of this book's authors, Pamela Matson (an ecosystem biogeochemist) along with Roz Naylor (an economist at Stanford University), and Ivan Ortiz-Monasterio (an agronomist at CIMMYT, the international maize and wheat research institute headquartered in Mexico); this case study is told in first person by P. Matson.

The Yaqui Valley is an interesting place for a couple of reasons. It includes approximately 250,000 hectares of irrigated agriculture (planted mostly in winter wheat) in the middle of the Sonoran Desert and receives its irrigation water from several huge reservoirs on watersheds that drain hundreds of thousands of acres in northern Mexico. It is located on the coast of the Gulf of California, one of the world's biodiversity hotspots. It is the birthplace of the Green Revolution, the place where Norman Borlaug and other researchers carried out early research and crop trials that led to the development of improved, high-yielding varieties of wheat, which then spread throughout the developing world during the Green Revolution years (1950–80s). (This early research was carried out in CIMMYT field stations, which remains today and is a focal area for continuing innovations and education.) The farmers in the valley had early access to the new, improved, higher-yielding varieties, and they received national subsidies to support the full complement of the Green Revolution technologies: fertilizer, irrigation, pesticides, and other inputs. As a result, farmers in the valley were very successful in increasing yields; they have some of the highest wheat yields in the world.

The Yaqui Valley is an excellent example of Green Revolution success; similar successes in many other parts of the world can be credited for food production keeping pace with the very rapid human population growth and growth in food consumption over the past 50 years. Unfortunately, however, the valley today has a range of sustainability challenges, most of them unintended consequences of the Green Revolution. For example, water resources are used very inefficiently (through flood irrigation in the fields), and until recently, the irrigation districts lacked rules for changing water draws to sustain water resources during times of drought (a situation that is not infrequent in the region). The agricultural systems tend to be overfertilized; the addition of far more nitrogen than is used by the crop leads to nutrient losses from soils to water and atmosphere, and most of the loss pathways have both economic and environmental costs. Shrimp aquaculture exploded unsustainably in the coastal zone of this region, which affected culturally important natural ecosystems and fisheries. Increases in winter temperatures related to climate change make some crops vulnerable to reduced or lost yields. Groundwater sources of drinking water are contaminated in some areas. Air pollution and pulmonary diseases in children are critical issues connected to agricultural practices.

Obviously, the Yaqui Valley is not on an optimal trajectory. While some might think it a good idea to completely convert this desert area into other nonagricultural activities, our research team was pragmatic. We sought to help move this region—a breadbasket of Mexico and a source of seeds and wheat for the world—toward sustainability. Although we made progress on several fronts, this short story focuses on fertilizer use.

In the Yaqui Valley, heavy subsidies precipitated the overuse of fertilizer. When industrial farming first started in the 1950s, very little fertilizer was applied (figure A.7). By 1981, farmers were applying plenty of fertilizer, enough to maximize the yields of the wheat they were growing then. Between 1981 and 1997 they continued to increase the application rates of fertilizer, even while yields stayed more or less constant. Moreover, farmers were applying most of that fertilizer prior to planting, followed by preplanting irrigation to sprout weeds and then, a month later, by seeding. Based on our own basic knowledge of biogeochemistry, we expected this approach could lead to big problems. Farmers, too, were beginning to be concerned about increasing fertilizer costs.

So, our research team decided to study the issue. We tracked the nitrogen fertilizer and tried to figure out where it was going, beyond the crop plants and soil. We found that nitrogen was being lost from these crop systems by a variety of pathways: into the atmosphere as the air pollutants nitric oxide and ammonia, which affect downwind air quality, including in urban areas; into the global atmosphere in the form of the long-lived greenhouse gas nitrous oxide; moving as nitrate ions through soil profiles into groundwater or surface water, with human health consequences; flowing out with the tail water of irrigation and into surface waters in the form of ammonium ions. All these

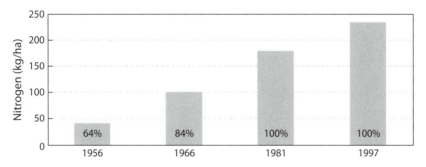

FIGURE A.7. Increasing annual rates of fertilizer application to wheat in the Yaqui Valley. Percentages represent the proportion of wheat area fertilized. (SARH—Mexico's Ministry of Agriculture and Water Resources, CIANO-Agricultural Research Center of the Northwest, and CIMMYT-International Maize and Wheat Improvement Center databases)

loss pathways were present in the system. Some of that nitrogen was then moving through surface and groundwater systems out into the Gulf of California, where it was causing huge phytoplankton blooms that swirled across the Gulf and into the protected marine systems on the other side. Some of the pollutants were spreading into the atmosphere, contributing to urban air pollution during parts of the year. Some of the nitrogen was probably being deposited downwind, affecting natural ecosystems there, although that wasn't something we were able to study. What was the bottom line? Fertilizer decisions that farmers were making in their fields were not sustainable: they were leading to strong levels of crop production and economic benefits for the farmers and others in the community, but they were also having unintended negative consequences on neighbors, ecosystems, water resources, and human health. Community voices were beginning to call for a change.

At the same time we were asking: Why are they doing what they're doing? Through surveys, interviews, and analysis of documents, we found that until the mid-1990s, fertilizers were incredibly cheap. Then, Mexico liberalized its agricultural policies, and fertilizer became expensive. By the late '90s, our analysis told us, it was the most important cost in farm budgets. But that was a recent change, and perhaps farmers hadn't caught up with the fact that the costs were so high. Farmers also said they were adding most of their fertilizer well before planting because of labor and machinery constraints (although a few farmers had overcome those constraints through machinery modifications). They also talked about their concern for getting enough fertilizer on the soils early, to optimize yields should they have a really good year in terms of weather, and to avoid potential rain events that would make the fields difficult to fertilize. But most important, they said their experience told them this management works.

Given the negative effects on natural and human capital both within and outside the valley's social-environmental system, we asked if there were some win-win opportunities for the farmers and the broader social-environmental system, something that could move them toward more sustainable cropping practices. To explore that question, we carried out field experiments (in farmers' fields and in agricultural experiment stations), developed and ran agronomic and biogeochemical models, and carried out more economic surveys and analyses. We subsequently found that, yes, there were some really great win-win opportunities. If farmers would apply less fertilizer, timing it more carefully to when the crop plants need it, they would be able to maintain yields, increase grain quality (the nitrogen content of grain), reduce all those nutrient-loss pathways tenfold or more, and save 12% to 18% of after-tax profit. That's a win-win!

We published a paper in *Science*[53] on our findings, and we suggested that farmers could be successful if they used this approach. But because our team cared about the well-being of the people in this place, and about sustainability goals there, we also actively worked to link that knowledge to what farmers were actually doing on the ground. We worked directly with farmers, doing on-farm trials of the new practices with many of the most innovative farmers in the valley, carrying out farmer workshops, and organizing field days and discussions groups. And we found that win-win options were working for those who tried it in their fields. We thus expected that the technology would spread throughout the valley.

But what actually happened? Again, because we cared about the sustainability of this place and not just about research publications, we carried out surveys a few years later to measure progress. We found that farmers, on average, were actually using *more* fertilizer, not less. Despite the win-win opportunity, and the economic benefits of using less, they were applying more fertilizer to each crop. Clearly, we did not understand something about this social-environmental system!

In the next segment of our research in the Yaqui Valley, we tried to understand exactly what was going on in the decision-making system. We studied the "knowledge system"—the network of actors and organizations that were producing, integrating, and using information in decision making. The knowledge system we thought we had when we started is shown in the upper left third of figure A.8. This system included researchers from universities working with key scientists like Ivan Ortiz-Monasterio in CIMMYT and others in the national extension groups, along with innovative farmers. But when we analyzed the knowledge system, we found something a lot more complicated than we had assumed was the case (see the entire figure A.8).

There are many important actors in this knowledge system, but the most critical players, we discovered, were the credit unions. These are farmer associations—respected and trusted organizations to which the farmers pay to belong. The credit unions provide credit for seeds and fertilizer and other inputs, access to and information about commodity markets, and advice about management. We discovered, through our analysis, that their advice had strings attached. We found that the credit unions were telling farmers, more or less: "if you want credit, apply high rates of fertilizer." So why was this the credit unions' advice? There is, of course, an economic incentive for a credit union to have its members buy agricultural inputs on credit, but there was another reason. The credit unions were aware farmers face a great deal of uncertainty; there is variability from farm to farm, from soil to soil, from farmer to farmer, from year to year. The credit unions' advice was designed to get around uncertainty and variability: if everybody puts on large amounts, everybody is going to do okay—that is, if one doesn't consider the externalities—the environmental and social costs related to overfertilization.

Given this reality, our research team recognized that something different was needed. We had provided a single "optimal" management practice, but we now realized that farmers and credit unions needed more field- and farmer-specific information. We needed to reduce uncertainty. In partnership with other research groups and farmers, we

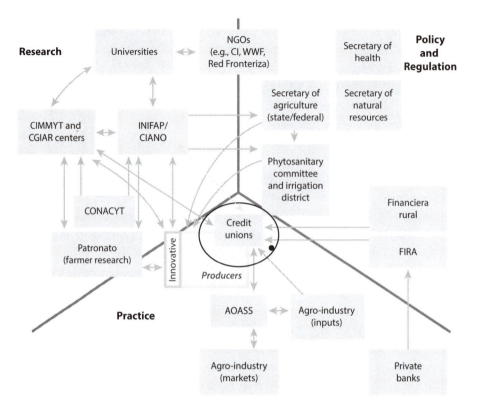

FIGURE A.8. The Yaqui Valley knowledge system. (McCullough, E., and P. Matson. 2012. "Linking Knowledge with Action for Sustainable Development: A Case Study of Change and Effectiveness." In *Seeds of Sustainability: Lessons from the Birthplace of the Green Revolution in Agriculture*, edited by P. Matson, 63–82. Washington, DC: Island Press)

developed a new technology—a handheld radiometer and associated data analytics—that could be used to tell farmers how much fertilizer their crop needed at any given time in a particular year on a particular soil, much more in tune with real-time decision making. And we worked collaboratively with farmers *and* the credit unions to develop and test this method. The credit union involvement was essential for several reasons, including that it was through them that the technology could be scaled up from just a few farmers to all the union members throughout the valley and beyond.

So how is it working today? In 2012, we carried out another survey and found that the new technology is continuing to spread. That's the good news. The bad news is that the change is very, very slow. And meanwhile, some farmers continue to use more fertilizer (perhaps encouraged by increasing wheat prices, especially for high-quality, high–nitrogen content grains). For rapid spread of this technology, we have to go beyond the win-win. Policies that tie best practices with access to farm subsidies and other agricultural support are used in the United States and other parts of the world; such regulations at the national scale may be the only way to move these changes forward at a rate and scale that can make a difference. Members of our team are engaged in this process now. Like most stories about transitioning to sustainability, this one is a continuing work in progress.

AN INTERNATIONAL SUCCESS AMID UNCERTAINTY: OZONE AND THE MONTREAL PROTOCOL

Invented in the mid-1800s, the mechanical refrigerator was a boon to human health and food quality. The ability to keep foods cool, even in the warmest days of summer, was viewed as a tremendous modern convenience that contributed to human well-being at numerous levels. Though iceboxes had been around for half a century and remained the most economical option for household use into the 1920s, mechanical refrigeration presented new possibilities for food storage and ice making. During the U.S. Civil War, for instance, one company used an ammonia-driven cooling machine imported from France to freeze water for New Orleans residents after ice supplies from the North were cut off.[54] Many grocers and shippers were also excited by the new technology.

However, the "fridge" was not without problems. Perhaps most important among those was the standard cooling fluid: ammonia, a gas with a dangerous propensity to explode. One highly publicized incident occurred when a refrigeration system at the 1893 Chicago World's Fair sparked a raging fire, and several firefighters lost their lives. Thus, not only were early refrigerators large, noisy, and expensive, but most individual households were afraid to purchase one because of their dismal safety record.

In the 1930s, during a search by industrial chemists to find a safer replacement for this dangerous cooling fluid, a new type of chemical was invented: a class of chemicals called chlorofluorocarbons, or CFCs. One of the most successful of these, marketed by DuPont under the trademark name Freon, very quickly replaced coolants in refrigerators and air conditioners. CFCs were later used in a variety of other applications, including as a propellant in aerosol sprays, as a cleanser for electronic components, and in plastics manufacturing. CFCs turned out to be remarkably useful for a variety of industries, and manufacturing quickly took off.

This new class of chemical was heralded as nontoxic and nonexplosive—and those claims were based on a thorough effort by DuPont scientists to check for both human toxicity and ecosystem damage. As far as any scientist could tell at the time, CFCs were a smart choice, both for people and the environment.

Connecting the Scientific Dots: A Slow Realization of Potential Harm

The potential for CFCs to play a more sinister role became apparent only later—in the 1970s and after—as atmospheric measurement technologies improved and the knowledge of atmospheric chemical interactions became more sophisticated. In 1970, Dutch scientist Paul J. Crutzen realized that human-made nitrogen oxide chemicals could react with and damage atmospheric ozone, but he did not immediately recognize the potential for CFCs to behave similarly.[55] Meanwhile, in the same year James Lovelock became the first person to measure very small quantities of CFCs in the atmosphere. He found them everywhere he looked, in the atmosphere over populated areas, as well as in regions far removed from industrial activities. Clearly, CFC gases were not very reactive in the lower atmosphere—and that meant that they stuck around and accumulated and got blown around the globe by large-scale wind patterns. While this was an interesting scientific finding, Lovelock had no evidence that it was a cause for concern.

Knowing that these unique human-made chemicals were accumulating in the atmosphere, albeit in very low quantities, other scientists began asking the question, what happens to them there? Along with Crutzen, Sherwood Roland and Mario Molina at the University of California–Irvine began to study the possible fates of the gases, out of scientific interest more than any expectation or concern that they could have negative environmental consequences. These scientists showed that the chemicals were basically inert in the lower atmosphere for decades, remaining unchanged by the visible wavelengths of light that penetrate to those atmospheric layers and also not reacting chemically with water and other molecules present in the lower atmosphere. They indeed were "safe" in that they were not involved in chemical reactions that could potentially create more harmful substances. However, the scientists' work also showed that in the upper atmosphere, or stratosphere, at altitudes of 30 kilometers or more, the CFC molecules are broken apart by the high-energy ultraviolet (UV) radiation from the sun there, forming much more chemically reactive fragments of chlorine atoms and other molecules of chlorine.

Based on chemical information stemming from laboratory chemical studies—not measurements high in the atmosphere—the scientists realized that these newly formed chlorine atoms can combine with (and thus remove) ozone—a gas molecule that protects Earth's surface from UV radiation. Moreover, they determined that when chlorine and ozone interact, they can cause a chain reaction that ultimately allows one chlorine atom to remove 100,000 molecules of ozone. Thus, in a paper published in 1974, Molina and Rowland suggested that the release of even small amounts of CFCs into the atmosphere could significantly reduce the protective layer of ozone in the upper atmosphere.[56]

This paper and related scientific work garnered attention. Since loss of upper atmosphere ozone would be expected to result in more UV radiation arriving at Earth's surface, the specter of increased exposure of people and other forms of life to UV radiation, accompanied by increased skin cancer, plant damage, cataracts, and other damages, was raised, and the ozone issue began to ring alarm bells. In the United States, the early 1970s was a time of growing public awareness and concern about environmental issues, and NGOs and the media brought this new concern to the public eye.

Meanwhile, the global scientific community (and especially the atmospheric sciences community) was intrigued, worried, and also questioning and challenging (as scientists typically are). Such an unexpected result, with such potential for societal consequences, not surprisingly engendered a surge in new research. The World Meteorological Organization (WMO) also expanded its activities related to research and monitoring. Meanwhile, international bodies like the Organization for Economic Cooperation and Development (OECD) and the United Nations Environment Programme (UNEP), along with the newly formed U.S. Environmental Protection Agency (EPA) began to focus attention on the potential regulatory issues, should the alarm turn out to be serious.

Taking Action amid Uncertainty

Although the alarm bells first rang in 1974, it took another thirteen years for governments of the world to agree on coordinated international action to reduce the risks of ozone loss. The fact that they finally did so is in many ways an amazing result. At the time, few thought that such an international effort could succeed. The story of the Vienna Convention for the Protection of the Ozone Layer in 1985 and signing of the Montreal Protocol in 1987 is a fascinating one. After many international data-sharing meetings, Finland and Sweden submitted the first draft of a potential treaty in 1982, later joined by Norway, Canada, and the United States in revised proposals (known as the Toronto Group).[57] This was in contrast to a European Community (EC) proposal that sought much less stringent controls on CFCs (at the time, Europe was the main exporter of CFCs globally).[58] Negotiations were conducted within an atmosphere of scientific complexity and uncertainty. Even the discoveries of what looked like an Antarctic "ozone hole" in the years 1984–85 by Japanese researcher Shigeru Chubachi and the British Antarctic Survey were not immediately seen as proof that CFCs *caused* the hole. Industry and civil society actors also wavered between supporting and opposing CFC regulations.

Despite the atmosphere of uncertainty, on September 16, 1987, twenty-four nations plus the European Community—the nations most responsible for the production and consumption of ozone-depleting gases—signed a treaty with the intent to reduce ozone-depleting substances by freezing the production of some and cutting in half the production of others. Among the unique characteristics of the treaty was the design feature of flexibility, allowing the timing and rate of reductions to change as the science changed. (And indeed, as the scientific evidence mounted and became much more alarming in the years following the Montreal Protocol, amendments to the protocol drove a much more rapid phaseout of CFCs and other ozone-depleting substances.) Further, the success in this case of the "convention-protocol" format, in which general principles are agreed to and then followed by more specific regulations, made it the gold standard for international environmental treaties.[59]

In the year immediately following the signing of the protocol, considerable debate still raged among some nations and corporations, but relatively quickly, industry responded. Aerosol replacements were already in use in the United States and quickly moved (in large part, due to public pressure) to Europe. Substitutes for CFCs used to clean electronic circuit boards, as refrigerants, and as propellants began to be tested for toxicity and moved into the marketplace. As Richard Benedict (who was the chief negotiator for the United States in this effort) said, "By providing CFC producers with the certainty that their sales were destined to decline, the protocol unleashed the creative energies and considerable resources of the private sector in a search for alternatives."[60]

Key Lessons in International Cooperation

Thanks to the flexibility of the Montreal Protocol, the story of ozone management continues. Among the many well-written descriptions of the Montreal Protocol process and outcomes, we find Richard Benedict's book, *Ozone Diplomacy*, to be an engaging read.[61] In it, Benedict gives an insider's perspective on the challenges and successes of the original treaty process, and what happened in the following years. While the story is fun to read, for our purposes we will share just some of the most critical take-home messages.

Benedict's book makes it clear that the success of the process that led to the Vienna Convention and Montreal Protocol rested on several key factors. First (and, he says, "foremost") was the role of science. This story is replete with scientific discovery, but Benedict notes that the scientific discoveries in themselves were not enough. More important, the scientific community had to be brought together to share learning, compare findings, settle scientific disputes, clarify uncertainties, and focus continued research on the most critical topics, including ozone chemistry and the state of the ozone layer. Moreover, scientists needed to share the knowledge on which they all agreed. Scientific assessments carried out by national and international bodies carefully articulated scientific understanding and presented it in readable reports that were essential to educating policy makers and the public. All these, and especially the broad-based assessment sponsored by the World Meteorological Organization in 1986,[62] were critical to the Montreal Protocol discussions. Likewise, the assessments that continued post-1987 were essential to the follow-up amendments that more rapidly limited the production of ozone-depleting substances. The bottom line is that scientists had to get out of the laboratory and out of their research institutions to help policy makers understand the issue and to communicate clearly with the public and corporate worlds.

Second, Benedict argues that a well-informed public was essential to mobilizing the political will of governments. Communication of findings by scientists, legislative hearings, an engaged press (including print and television), and government as well as non-governmental education campaigns all contributed to informing citizens around the world. In the United States, consumers were already asking questions about the environmental consequences of CFCs by the mid-1970s, and as a result, their use as propellants in aerosol cans was banned by the end of the decade. While the uncertainties of ozone science (and the lack of a signal of ozone loss until the 1980s) at times confused the public (and was used to engender confusion by some industry members and government leaders), the ongoing engagement of the public in this issue ultimately spurred action.

Third, Benedict argues that UNEP, a multilateral institution created in 1972, played an essential role throughout the process. Under the leadership of Mostafa Tolba of Egypt, it mobilized science across countries and shared information, informed public opinion, cajoled and pressured governments to the bargaining table, and in a sense, became a voice for developing countries not present at the negotiations.

The fourth critical factor was the individual nations' policies and leadership. The U.S. government had a large and relatively well-funded scientific community that contributed scientific knowledge, along with important contributions from researchers worldwide. The governments of the United States, the Nordic countries, and Canada showed deep commitments early on and enacted some of the earliest legislation against CFCs used in aerosol can propellants. National leaders in a number of countries—including the United States, Canada, Japan, Australia, Mexico, New Zealand, and the former Soviet Union—encouraged government-wide engagement in scientific, technological, and economic fact-finding, as well as in the international policy process.

Finally, Benedict notes that non-governmental organizations (NGOs), including both industry and citizen groups, participated actively, informing the public, pressuring governments, and offering their perspectives to negotiators and the press. In the United States, industrial producers had to contend with the strength of public concern and the threat of a patchwork of state regulations as well as early EPA regulation of propellants. At the same time, they were concerned about the advantage of European industries that were dominating part of the market. As a result, U.S. industry was more supportive of international action than European companies, who clearly wanted to preserve their market dominance of CFCs and avoid costs related to switching to new technologies. In 1986, a coalition of 500 U.S. producers and users of CFCs and other ozone-depleting substances issued a statement to support international regulation.

The design of the process itself, according to Benedict, also was critical to the outcome. With the help of nations and UNEP, the complex challenge was broken down into pieces, with prenegotiation groundwork carried out via scientific and economic workshops and brainstorming sessions, and many bilateral discussions.

Finally, and perhaps most important for its long-term impact, the Montreal Protocol was designed to learn. It was a flexible instrument that could adapt based on periodic scientific reassessments as well as economic and technological updates. At the time of the Montreal Protocol, no one had yet demonstrated a significant reduction in total ozone (that record of loss was to become clear only in the following two decades), and the relationship of CFCs to the newly identified ozone hole over the Antarctic was still under study. There was little contention (at least among the negotiators) that changes in the ozone layer would pose risks for humanity, but there was still considerable uncertainty about how much it would change. Given the fact that science—especially the science of complex social-environment systems—is always uncertain, and that new information would continue to be forthcoming through research, the protocol benefited dramatically from being an ongoing process rather than a static solution. Indeed, the Montreal Protocol has been updated multiple times with adjustments (for urgent updates) and amendments (for more comprehensive changes).[63]

What is the bottom line for this case study? Clearly, the CFCs were a useful technology that helped meet some of the needs and wants of people around the world. As with most technological innovations, however, they had unexpected consequences beyond solving the problem for which they were targeted—consequences that science at the time did not recognize. Thanks to curiosity-driven discovery research as well as use-inspired, problem-solving research; good environmental measurement and monitoring systems; engaged decision makers in governments, industry, and civil society; and a good bit of luck, the ozone problem was identified and addressed. While the loss of stratospheric ozone has not been completely reversed, the decline has clearly been arrested as a result of concerted social action, and signs of a recovery are evident.

Glossary of Terms, Acronyms, and Additional Resources

GLOSSARY OF TERMS

Action   **Action** includes the behavior, decisions, and day-to-day practices that people choose to pursue (chapter 4).

Actors   **Actors** are the people, groups, or organizations that are involved in the decision-making process or that will be affected by the decisions. Actors can be as large as countries or multinational organizations and as small as households or individual people. They have particular characteristics that include values, beliefs, power, agenda, interests, capacity, and motivation (chapters 2 and 4).

Adaptation   **Adaptation** refers generally to a learning cycle. It takes place throughout innovation cycles as initial knowledge contributions are modified to better fit the needs of particular users, to accommodate characteristics of the production process, or to take advantage of new discoveries and learning. Adaptation is hampered by obstacles such as willful ignorance, the inability to admit error, and the lack of shared forums for learning (chapter 5).

Adaptive Management   **Adaptive management** is an active arena of research and practice. It treats policy and management regimes as experiments to be learned from rather than as final "best bets." Its emphasis is on careful design of interventions to maximize what can be learned from them, implementation of appropriate monitoring regimes to detect failures early, and timely retirement of those failures to make room for alternative interventions (chapter 5).

Agency   **Agency** is the degree to which a given actor has the capacity to act independently and to make his or her own decisions. While agency is an abstract concept that is difficult to measure directly, it is possible to characterize actors' degree of agency relative to that of other actors by comparing their access to political power, financial resources, and information (chapters 2 and 4).

Assets. *See* Capital Assets.

Boundary Work **Boundary work** signifies the processes through which the research community organizes its relations with the worlds of action and policy making with practice-based and other forms of knowledge. The idea of boundary work has been applied to the interface between science and policy and, more broadly, to the activities of organizations that seek to mediate between knowledge and action. The central idea of boundary work is that tensions arise at the interface between actors with different views of what constitutes reliable or useful knowledge, and those tensions must be managed effectively if the potential benefits of research-based knowledge are to be realized by society (chapter 5).

Capabilities **Capability** is the power or ability to do something. People's capabilities are related to their levels of experience, education, and practical knowledge, in addition to the opportunities afforded them by society or governments (chapter 2).

Capital Assets **Capital assets** constitute the ultimate determinants of well-being—the stocks of capital on which society draws to create its well-being. The framework presented in this book characterizes a society's assets in terms of five component capital assets: natural, manufactured, human, social, and knowledge (chapter 2 and figure 2.1).

Collective Action **Collective action** occurs when two or more actors cooperate to accomplish a goal they cannot achieve individually. For example, a group of farmers build and maintain an irrigation system, or a group of students organize a campaign to protest against their university's refusal to adopt a zero-carbon emission goal. Many of the traits of a functioning society—such as law and order, public safety, economic growth, and environmental protection—require coordinated, collective actions at a very large scale (chapter 4).

Collective-Action Problem A **collective-action problem** occurs when actors fail to accomplish their common goal. The creation of governance arrangements faces potential collective-action problems. The chances of overcoming these problems are greatly enhanced by an appreciation of the factors that motivate different governance actors to come together and actively contribute to governance responses that represent the common good (chapter 4).

Common-Pool Resource Many of our natural resources such as forests, groundwater, and fisheries may be considered **common-pool resources (CPRs)**. There are two defining characteristics of CPRs: they are finite resources, and it is difficult to prevent anyone from extracting from them. Because of these attributes, a rule system that is recognized by resource users is needed to control access to and regulate use of the resource (chapter 4).

Community **Community** is the totality of social system interactions within a defined geographic space. Communities have some level of "shared fate" (e.g., they might all experience a hurricane or earthquake) and comprise unique combinations of manufactured, social, human, and natural capitals (chapter 2).

Complex Adaptive Systems **Complex adaptive systems** are systems that have multiple interconnected components with feedbacks and time lags affecting their interactions, exhibit nonlinearities and tipping points that influence the ways they respond to interventions, and are characterized by self-organization and emergent behavior (in which the system as a whole is more complex and organized than would be predicted by the behavior of the individual parts) (chapter 3).

Connectivity In general, connectivity refers to the state or extent of being linked or connected. More specifically for our purposes, **connectivity** refers to the ways in which resources, species, decision makers, and other components of social-environmental systems interact and spread across ecological or social landscapes (chapter 3).

Constituents of Well-Being Constituents are parts, components, or dimensions. The constituents of a person's appearance thus include their face, body shape, clothing, and the like. The **constituents of well-being** are the dimensions of the "good life" that people feel are most important to them. These will differ from individual to individual. For most people, however, the important constituents of well-being include the capacity to meet their basic needs for food, water, shelter, energy, and physical security. Many would add to this list of basic needs additional factors that make life not just livable but good: access to health, education, nature, a sense of belonging, and the capacity and opportunity to shape one's own life (chapter 2). *Contrast* Determinants of Well-Being.

Consumption Processes   **Consumption processes** can be defined as the human use of goods and services created by production processes to achieve (or not) the goals of sustainable development (i.e., in our case, inclusive social well-being). In the context of sustainability, consumption is important in that it is the process by which many needs are met, but it is also a process that may make materials or energy less available for future use, may alter an ecological system so that it provides less services now and in the future, or may threaten human health, welfare, or other things people value (chapter 2).

Credibility   **Credibility** is a perception of knowledge. In particular, it concerns perceptions of the likely truth of knowledge: Does the potential user of new knowledge have reason to believe that the person or organization providing that new knowledge actually knows what she or he is talking about? For the purpose of linking knowledge to action, credibility must be constructed for particular users in particular contexts using standards that work for them. Together with the properties of saliency and legitimacy, credibility produces trust (chapter 5).

Determinants of Well-Being   Determinants are ultimate causes. The determinants of a person's appearance are thus their genetics, age, nutritional history, and the like. The **determinants of well-being** are the stocks of capital assets that fuel sustainable development: natural capital, human capital, manufactured capital, social capital, and knowledge capital (chapter 2). *Contrast* Constituents of Well-Being.

Development   According to the Bruntland Commission, "**Development** is what we all do in attempting to improve our lot" within our environment. We use this term without a modifier to mean simultaneous economic, human, and social development (chapters 1 and 2).

Education   **Education**, the knowledge gained from systematic instruction, is the foundation for individual self-improvement as well as for our collective advancement as humans. Education is what allows people to take advantage of the reservoir of knowledge capital in the world (chapter 2).

Ecosystem Services   **Ecosystem services** refer to those benefits to humans that arise from the function of ecosystems in our environment (for example, water filtration and storage, the uptake and release of CO_2 from the atmosphere by ecosystems, food and fiber production, erosion control). Ecosystem services are technically a subset of environmental services, but the two terms are often used interchangeably.

Environmental Services. *See* Ecosystem Services.

Evaluation   **Evaluation** is a central concept in sustainability science because it is a use-driven pursuit. Thus, in the name of sustainable development, we seek to evaluate whether we have been developing sustainably or whether a given future policy or trajectory will promote sustainable development. We use the term as a broad one, encompassing what others call *assessment, analysis,* and *reporting* (chapter 1).

External Forces   **External forces** are exerted on a system from outside system boundaries. At all scales of a social-environmental system, and all levels of governance within a system, there are factors beyond the control of the actors at any given level (e.g., household, community, or region). Examples of external forces include solar radiation, climate change, global population growth, natural disasters, global price fluctuations, decisions higher up in the political hierarchy, and international agreements (chapters 3 and 4).

Externalities   **Externalities** are generally understood to be side effects of an activity that affect other parties but are not reflected in the cost of the goods or services rendered by the activity. Sometimes, individual decisions that are made in one social-environmental system have significant consequences for people and systems not directly involved with the decisions. When these consequences cause damage, they are referred to as **negative externalities**, because decisions adversely affect third parties who are external to the setting in which the decision was made (for example, discharges of waste into a river). **Positive externalities** also exist. The form most relevant to sustainable development may be technical innovation. Inventors often see the fruits of their work adopted without payment by others who did not bear the costs and risks of research and experimentation. Externalities are examples of invisibilities that make the management of social-environmental systems so challenging (chapters 3 and 4).

Feedback Loop **Feedbacks** result when changes in one part or component of a system—a particular process or variable—cause changes that eventually loop back to influence that component, either by reinforcing the change in that component (a positive feedback) or dampening the change (a negative feedback). Feedbacks result from a chain of events that can sometimes be very hard to measure or predict (chapters 3 and 4).

Free-Ride (Free-Riding) To **free-ride** is to benefit from a shared good or service without contributing to its provision. It is a common barrier to cooperation (chapter 4).

Governance The concept of **governance** includes both rules and the processes of generating and using those rules. In the context of the framework for understanding sustainable development that we use in this book (figure 2.1), the "rules" of governance are an important part of a society's assets of social capital. These rules, which we call **institutional arrangements**, reflect social agreements on what people may, must, or must not do in their interactions with one another and with their capital assets. They also define who has the rights and responsibilities to monitor and enforce rules. Finally, they specify who has the right to participate in **governance processes** that lay down new rules and revise existing ones, providing a means through which the nature of social-environmental interactions can be changed so that human activities do not drive down the overall asset base on which future generations depend. Social actors and agents of change who agree to create, implement, and abide by rules are engaged in processes of governance. Governance is important in the pursuit of sustainability in that its rules or arrangements are among the core components of social capital and thus are among the determinants of inclusive wealth and well-being (chapter 4). *See also* Governance Processes and Institutional Arrangements.

Governance Processes **Governance processes** lay down new rules and revise existing ones, providing a means through which the nature of social-environmental interactions can be changed so that human activities do not drive down the overall asset base on which future generations depend. Social actors and agents of change who agree to create, implement, and abide by rules are engaged in processes of governance (chapter 4). *See also* Governance and Institutional Arrangements.

Health **Health**, an expression of a person's mental and physical condition, is probably the most universal and immediately recognizable component of well-being. Without good health, an individual's quality of life suffers, making it more difficult for individuals and communities to reach their potential or other dimensions of well-being. Healthy people help make up strong human capital and are thus a determinant as well as a constituent of well-being (chapter 2).

Human Capital **Human capital** is the stock of a system's productive assets embodied in individual people. We focus on three components that together determine the quantity and quality of human capital: (1) the size, age structure, and geographic distribution of the human population; (2) the health of that population; and (3) the acquired capabilities (education, experience, tacit knowledge) of the people who constitute that population. It is helpful to think of high human capital as "healthy, well-educated, skilled, innovative and creative people," and a key task of sustainable development is figuring out how to foster these characteristics (chapter 2).

Human-Environment Systems. *See* Social-Environmental Systems.

Incentives The term **incentives** refers to things that motivate actors to do something—the expected rewards and punishments that individuals perceive to be related to their actions and those of others. Rewards can be monetary, such as a wage, tax credit, rebate, or a bonus. Rewards can also be nonmonetary in nature and include earning the respect of one's peers, the satisfaction from learning new skills or knowledge, affection, and the feeling of having done the right thing. The threat of punishments of different kinds can also function as incentives. Such punishments may be the threat of a lawsuit, consumer boycott, fine, jail sentence, social exclusion, or loss of employment. This term is closely associated with the concept of governance, through which incentives are generated that motivate individuals to make decisions that overcome collective-action problems and externalities (chapter 4).

Inclusive Social Well-Being **Inclusive social well-being** is well-being of individual people aggregated across space and time. We define development to be sustainable over places and times in which inclusive social well-being does not decline. We often use the shorthand "inclusive well-being" or just "well-being" (chapter 2).

Information Problems **Information problems**, with respect to collective action, exist in situations in which actors do not share the same information about available choices, the consequences of those choices, or the characteristics and preferences of other actors with whom they interact (chapter 4).

Innovation We adopt the conceptualization due to Harvey Brooks, who characterized **innovation** as "the process by which technology is conceived, developed, codified, and deployed." This definition is not as narrow as it looks, for Brooks held a broad view of "technology," portraying it as "knowledge of how to fulfill certain human purposes in a specifiable and reproducible way." Technology, and thus innovation, therefore includes not only devices but also policies and procedures (Brooks, H. 1980. "Technology, Evolution, and Purpose." *Daedalus* 109[1]: 65–81).

Institutional Arrangements **Institutional arrangements**, as we use the term here, are a society's "rules of the game" that influence how people interact with one another and the rest of the social-environmental system. Those rules can be formal or informal. Examples of rules include policy, regulations, local norms and customs, contracts, and property-rights arrangements. Rules specify not only the rights and responsibilities for using and managing social-environmental systems but also whose responsibility it is to monitor and enforce rule compliance (chapter 4).

Institutions. *See* Institutional Arrangements.

Interventions We use **interventions** as a general term encompassing all the things that people could do to change the path of development. This includes the introduction of new or existing but not utilized policies, technologies, tools and information. We distinguish such interventions from the narrower idea of **innovations**, which we use to signify the creation of new technologies and policies (chapter 1).

Invisibilities To be invisible is to be out of sight, concealed, or hidden. In social-environmental systems, **invisibilities** refer to the linkages between systems across space and time that are not obvious (e.g., the movement of pollutants in air or water) and the fact that decisions made in one place or at one point in time can have many consequences in places and times far distant, thereby affecting people and social-environmental systems not engaged in the decision making (chapter 3). *See also* Externalities.

Knowledge Capital **Knowledge capital** encompasses both conceptual and practical knowledge, including general principles, information, facts, devices, and procedures that are intangible public goods; they can in principle be used by anyone who wishes to do so, and they can be used repeatedly, without depletion (chapter 2).

Learning **Learning** can be usefully thought of as the process of acquiring new or modifying existing knowledge, skills, and behaviors. It is of fundamental importance in building resilient systems (chapter 3).

Legitimacy **Legitimacy**, for our purposes, is the most subtle of the dimensions of trust in knowledge. Legitimacy is about perceptions of fairness, lack of bias, and respect. Other dimensions of knowledge that contribute to trust are credibility and saliency (chapter 5).

Life Support System The **life support system** of Earth includes the interactions among its atmosphere and climate; its processing of key materials such as oxygen, carbon, nitrogen, phosphorus, and water; its ecosystems and biodiversity; and the soil, mineral, and other resources that together are essential for life.

Management In the context of social-environmental systems, **management** is the day-to-day activities that people undertake in direct interactions with the capital assets. In that sense, management is an operational, field-level activity, whose purpose is to alter the productivity of a particular aspect of the capital assets. An example related to natural capital would be the pruning of trees or selective cutting for the purpose of improving the commercial profitability of timber logging (chapter 4).

Manufactured Capital **Manufactured capital** (which some call "produced capital") includes human-made factories, transportation systems, dwellings, and energy infrastructure as well as the objects—from books and artwork to shoes and blankets—that enrich our daily lives. Despite providing many benefits to human well-being, manufactured capital can also cause damage to other forms of capital in the productive base of the social-environmental system and thus challenge the goal of sustainable development. For example, destruction of forests to provide resources for manufacturing can cause a range of ecosystem services to decline (chapter 2).

Material Needs. *See* Needs.

Models **Models** are simplified representations of reality or idealized systems (i.e., scenarios); they can be conceptual or mathematically formalized; and they are used in a variety of ways in sustainability science (chapter 3).

Motivation One of the most common sources of breakdowns in collective action is related to weak individual **motivation** (desire) to contribute to the common good. The better off an individual is in the status quo, the weaker his or her motivation to invest in a process that might change the status quo. Even if an individual would like to see a change in the status quo, that person is tempted to let others do the heavy lifting in the change process (chapter 4).

Natural Capital **Natural capital** is the entire "Earth system" that provides for the basic needs of all people. It includes the atmosphere and climate, mineral resources, ecosystems and biodiversity, biogeochemical cycling, soils for growing crops, the crop plants themselves, high-quality groundwater or surface water resources, materials to build with, the ocean fisheries, and many other sources of the goods and services needed by humans (chapter 2).

Needs **Needs** (essentials or necessities) can be difficult to define, as people have different notions of needs (versus wants or luxuries), but at the most basic level, humans need food, water, and shelter to survive. When these **material needs** are satisfied, they provide a foundation of well-being from which to engage in personal or professional growth. With sustainable development, short-term needs do not trump the needs of people in the longer term; intragenerational well-being does not trump intergenerational well-being. The needs of one person or community do not trump needs of others in other places (chapter 2).

Opportunity **Opportunity**, the set of circumstances that makes it possible to do something, enables people to make choices about how they want to live and what they want to do. Opportunity is a key constituent of well-being because of its implications for experiencing "real freedom," or the capacity to pursue an unrestricted range of choices in life given one's personal capabilities (chapter 2).

Political Process   The **political process** focuses on the ways in which identified actors interact with one another to make collective decisions. The political process can exhibit varying degrees of transparency, accountability, representation, and legitimacy (chapter 4).

Power Asymmetries   **Asymmetries** (inequalities) **of power** exist when some actors are more powerful than others. The basis for holding power can be economic, social, or political. When actors do not have a say, or even a seat, at the decision-making table, they lack power in the governance process. When this occurs there is a risk that the needs of marginalized groups are not taken into account when collective decisions are made. Future generations also represent a group whose needs may not be taken into account in governance decisions (chapter 4).

Production Processes   **Production processes** transform capital assets into goods and services that people can use or consume. Production of food, electricity, clothing, housing, manufactured goods, along with their accompanying wastes, are examples (chapter 2).

Productive Assets. *See* Capital Assets.

Regime Shift   A **regime shift** is a large, persistent, and often abrupt change in a system that occurs because of substantial changes in feedbacks and other interactions and forcings in the system (chapter 3).

Resilience   **Resilience** is the ability of a social-environmental system to continue to function and support its current mode of operation despite stresses, challenges, and external forces that attempt to make it do otherwise. While there is no definitive set of characteristic features of resilience, those commonly discussed are diversity, redundancy, and connectivity, which, in the context of social-environmental systems, include both the social and environmental components of the system (chapter 3).

Rules   **Rules** help organize the way social-environmental systems interact. Individuals who agree to create and implement rules are engaged in a process of governance (chapter 4).

Saliency **Saliency** (prominence or importance) is about perceptions of the relevance of knowledge: do potential users see the expert advice or new technologies they are being offered as bearing on their real needs? Reorienting knowledge production efforts to assure that they are (and appear to be) salient to the most intensely felt needs of potential users is an essential step to building trusted and influential knowledge (chapter 5).

Security The United Nations Development Programme defines human **security** as the "freedom from want and fear." In this sense, human security encompasses more than just peace between nations; it includes individual freedom from crime, discrimination, and hunger (chapter 2).

Social Capital **Social Capital** includes the economic, political, and social arrangements—including laws, rules, norms, networks, institutions, and trust—that influence how people interact with one another, the environment, and other components of social-environmental systems (chapter 2).

Social-Ecological System. *See* Social-Environmental System.

Social-Environmental System A **social-environmental system** includes people, their institutions, technologies and manufactured goods, and the planetary resources and environment, all interacting in a tightly coupled way. Social-environmental systems are complex systems with many elements or components that interact in diverse ways, and with positive and negative feedbacks, invisible connections across space and time, and nonlinearities and tipping points that influence the way the system works and the way it changes with each intervention. Such systems are also referred to as **human-environment systems** or **social-ecological systems** (chapter 3).

Sustainable Development Modern usage is anchored in the definition provided by the World Commission on Environment and Development (the Brundtland Commission) in 1987: development is sustainable if "it meets the needs of the present without compromising the ability of future generations to meet their own needs." In this book we expand the spare "needs" focus of the Brundtland Commission to embrace the broader concept of "well-being." We retain the view of the commission that development should involve all members of society now and in the future, and capture these ideas in the phrase "inclusive social well-being." In our usage, **development is sustainable** if inclusive social well-being does not decline over multiple generations. This is the same definition we adopt for "sustainability." We therefore use the two terms interchangeably (chapter 1).

Sustainability   **Sustainability** is a term in common use with a variety of implicit meanings. Common to most is the realization that our ability to prosper now and in the future requires increased attention not just to economic and social progress but also to conserving the "life support systems"—the fundamental environmental and resource assets—on which our hopes for prosperity depend. As used in this book, sustainability is achieved to the extent that inclusive social well-being does not decline over multiple generations. This is the same definition we adopt for "sustainable development." We therefore use the two terms interchangeably (chapter 1).

Sustainability Science   **Sustainability science** is a field that, like the fields of health or agricultural science before it, is all about "use-inspired research," creating knowledge to help address social problems. In particular, it is a field that integrates study and practice, focuses on the interactions between environment and development, and performs use-inspired research to promote the goals of sustainability (chapters 1 and 6).

Sustainability Transition   Sustainable development is not a destination but, rather, a goal-driven process. A **sustainability transition** involves altering those processes so as to "bend the curve" from its present unsustainable trajectory toward one more consistent with achieving sustainability goals (chapters 1 and 6).

System   A **system** is a bounded area with a set of elements or components that are connected to and interact with one another. An ecosystem, for example, whether a forest or lake or grassland or coral reef or agricultural field, consists of all the organisms (plants, animals—including humans, and microbes) and the soil, water, air, rocks, chemicals, and the like, with which they interact, and it can include the people who use or manage the ecosystem as well. **Social-environmental systems** are of particular interest for sustainability (chapter 3).

System Flows In any system, the size or amount of the stock is controlled by inputs and outputs (or inflows or outflows). For example, the amount of water in a bathtub at a particular point in time (the stock of water), and its change over time, is a function of how much water is flowing in minus how much water is flowing out (chapter 3).

System Stocks **System stocks** are the amounts of particular components in a system (for example, the amount of biomass in the trees of a forest, the amount of water in a storage tank or aquifer, the number of people in an office building or city) (chapter 3).

Technology We adopt Harvey Brooks's broad view of technology as "knowledge of how to fulfill certain human purposes in a specifiable and reproducible way." Technology therefore includes not only devices but also policies and procedures (Brooks, H. 1980. "Technology, Evolution, and Purpose." *Daedalus* 109[1]: 65–81).

Tipping Point A **tipping point** is a critical threshold at which even a small perturbation can alter the state or fundamental way a system functions, leading to a regime shift (chapter 3).

Trust We use **trust** in the broad sense "of someone entertaining correct expectations regarding someone else's promises . . . including cases where the promises are implicit by virtue of the community's customary norms" (Dasgupta, P. 2010. "A Matter of Trust: Social Capital and Economic Development." In *Lessons from East Asia and the Global Financial Crisis*, J. Y. Lin and B. Pleskovic, eds., 119–55. Annual World Bank Conference on Development Economics Global. Washington, DC: World Bank) (chapters 2 and 5).

Use-Inspired Research **Use-inspired research** is research that contributes to discovery (basic research) and to problem solving (applied research) and facilitates interaction between these two forms of research (chapters 1 and 6).

Vulnerability **Vulnerability** is the likelihood of suffering harm. In the context of social-environmental systems and sustainable development, vulnerability refers to a system or components of the systems (e.g., groups of people, infrastructure, ecosystems, water resources) that will most likely suffer damage owing to disturbances, with subsequent impact on inclusive well-being (chapter 3).

Well-Being **Well-being** is the state of being comfortable, healthy, and secure because of having basic needs met as well as having access to health, education, community, and opportunity. The core of the human experience of well-being rests in a combination of material, social, and personal fulfillment (chapters 1 and 2). *See also* Constituents of Well-Being.

ACRONYMS

ASB	Alternatives to Slash and Burn
CFCs	Chlorofluorocarbons
CPR	Common-pool resource
EC	European Community
EPA	U.S. Environmental Protection Agency
GDP	Gross domestic product
GED	Gross external damage
GIS	Geographic information systems
GNI	Gross national income
HDI	Human Development Index
IWI	Inclusive Wealth Index
LCA	Life cycle assessment
MCA	Multi-criteria analysis
NASA	National Aeronautics and Space Administration
NGO	Non-governmental organization
NOAA	National Oceanic and Atmospheric Administration
OECD	Organization for Economic Cooperation and Development
RCT	Randomized controlled trials
RISA	Regional Integrated Sciences and Assessment
UNDP	United Nations Development Programme
UNEP	United Nations Environment Programme
UV	Ultraviolet
WCED	World Commission on Environment and Development
WMO	World Meteorological Organization
WWF	World Wildlife Fund

ADDITIONAL RESOURCES

Chapter 1. Pursuing Sustainability: An Introduction

The following are several broad perspectives on sustainable development that are interestingly different from ours.

1. World Commission on Environment and Development (The Brundtland Commission). 1987. *Our Common Future.* New York: United Nations.

   This report from a generation ago launched the modern era of global thinking about sustainable development. It's clearly dated but remarkable reading nonetheless, reflecting the commission's unparalleled efforts to visit multiple places around the world collecting data and testimony to inform its efforts. See especially the chairman's foreword. Available through the following link for online reading or free download: www.un-documents.net/wced-ocf .htm.

2. Sachs, J. 2015. *The Age of Sustainable Development.* New York: Columbia University Press. 544 pages.

   This encyclopedic coverage of the field is from one of the leaders of the international effort to create a set of formal sustainable development goals (SDGs) for the post-2015 era. It covers more substantial topics than ours. Individual chapters discuss sustainability in the context of energy, agriculture, urban, and other development efforts, as well as the risks that such development poses to climate change and biodiversity loss. The price of this comprehensive approach to the world as it is may be less of an orientation to the long-term future and a less consistent analytic framework for thinking about sustainability than the shorter work we present here.

3. Adams, W. M. 2008. *Green Development: Environment and Sustainability in the Third World,* 3rd ed. London: Routledge. 480 pages.

   This text has had more staying power in the field than perhaps any book other than the original Brundtland Report. One of the reasons is that it is so thoroughly grounded in the author's extensive experience in grappling with the interactions between development and environment in some of the poorest but also most rapidly growing parts of the world. He provides a discussion of issues particularly relevant to those regions—drylands, dams, biodiversity conservation, and the like. Perhaps most important, he offers one of the most balanced assessments we know of alternatives to "mainstream" thinking about sustainability (including ours)—for example, ecofeminism, deep ecology, ecosocialism, and political ecology.

4. Kates, R. W., ed. 2010. *Readings in Sustainability Science and Technology*. CID Working Paper No. 213. Center for International Development, Harvard University. Cambridge, MA: Harvard University, December 2010.

Presidential Science Medalist and geographer Robert Kates has prepared an annotated "reader" of the classic journal publications and reports in the field. It's a delight to browse, both for the author's comments accompanying the recommended readings and for his effort to provide a structure with which to make sense of the burgeoning literature in the field.

5. Graedel, T. E., and E. van der Voet, eds. 2010. *Linkages of Sustainability*. Cambridge, MA: MIT Press. 552 pages.

This book explores how different development sectors interact via the common demands they place on key material and energy resources (e.g., competition for water of the energy and agricultural sectors). It brings together an international cross section of authors to identify where such common demands may pose systems constraints to sustainable development and to assess options for surmounting them.

Chapter 2. A Framework for Sustainability Analysis: Linking Ultimate Goals with Their Underlying Determinants

1. Dasgupta, P. 2004. *Human Well-Being and Natural Environment*. New York: Oxford University Press. 376 pages.

This work is the most complete development of the theory behind the framework presented in our book. It presents the argument that sustainable development should be defined in terms of human well-being and argues that the determinants of sustainable development are society's capital assets. It is especially sensitive to the fact that for poor people in the developing world, natural capital often constitutes their greatest asset, yet that asset is systematically undervalued—and thus often degraded—by conventional development models. The author's core arguments are given a formal mathematical treatment, though the text remains accessible to those for whom the math is not.

2. Sen, A. 2013. "The Ends and Means of Sustainability." *Journal of Human Development and Capabilities* 14 (1): 6–20.

This essay, originally published in 2000, argues for a broad conception of human well-being that includes people's capabilities to shape their own lives. It also makes a strong case for the importance of distinguishing our ultimate ends of (goals for) sustainable development from the means by which we hope to attain those ends. The author calls for a partnership of scholars and leaders to provide the "informed agitation" needed for the pursuit of sustainability.

3. World Bank. 2010. *The Changing Wealth of Nations: Measuring Sustainable Development in the New Millennium*. Washington, DC: World Bank. 221 pages.

Economists at the World Bank played a central role in the development of the assets-based perspectives on sustainability that we discuss in this book. Here is the second of the bank's comprehensive efforts to define its approach. This volume explores the implications of moving beyond GDP to incorporate the value of natural capital in reassessing development progress at the national scale.

4. UNU-IHDP and UNEP. 2012 (2014). *Inclusive Wealth Report 2012 (2014): Measuring progress toward sustainability*. Cambridge: Cambridge University Press. 336 pages.

This report, produced by a collaboration of international organizations, builds on the World Bank effort cited above. It provides measures of the capital assets discussed in our framework and aggregates them to estimate progress toward sustainable development for most countries of the world. Accompanying essays outline the theory behind the measures and some of the central challenges of improving our ability to track sustainable development.

Chapter 3. Dynamics of Social-Environmental Systems

There is no one perfect text or manuscript that covers all you might want to know about coupled social-environmental systems. We provide some good starting places here. We also suggest that one excellent way to explore the full suite of challenges in social-environmental systems is to read the *Annual Review of Environment and Resources*. This annual publication provides, over a rolling five-year period, an up-to-date review and syntheses of the full range of issues central to the sustainable development of social-environmental systems. We strongly recommend it!

1. Sayer, J., and B. M. Campbell. 2004. *The Science of Sustainable Development: Local Livelihoods and the Global Environment.* Cambridge: Cambridge University Press. 292 pages.

   The authors of this book have many years of experience in addressing sustainable development problems in developing countries through problem-oriented, collaborative, and interdisciplinary research activities. The work they do is an example of boundary-spanning work, in which decision makers at different levels play a prominent role in the research process. Their book calls for researchers to reform our present research system so that it becomes more effective in achieving what they call "integrated natural resource management" (a reconciliation of development and conservation objectives). The book presents several case studies from the developing world that are used to illustrate the arguments about the role research ought to play in moving toward better integration of conservation and development goals. The new research approach they advocate for is similar to what we discuss in our book in that Sayer and Campbell describe how a variety of scholarly perspectives are useful for advancing the pursuit of sustainable development.

2. De Vries, B. 2013. *Sustainability Science.* Cambridge: Cambridge University Press. 605 pages.

   This is the best comprehensive textbook we know of in the science of sustainability. It is targeted at master's students, but many portions could be used in courses at the advanced undergraduate level and could serve as a desk reference for research students. It uses systems dynamics as its organizing principle but also provides accessible qualitative discussions of human–environment interactions in past civilizations and in contemporary development sectors.

3. Lee, K. N., W. Freudenburg, and R. Howarth. 2012. *Humans in the Landscape: An Introduction to Environmental Studies*. New York: W. W. Norton & Company. 431 pages.

    This textbook links the scientific analysis of the traditional field of environmental studies with the approaches and perspectives of sustainable development. Written for undergraduate classes, it discusses current grand challenges of environment, the emergence of environmental problems, and strategies being employed to address them.

4. Komiyama, H., K. Takeuchi, H. Shiroyama, and T. Mino. 2011. *Sustainability Science: A Multidisciplinary Approach*. Tokyo: United Nations University Press. 375 pages.

    This work traces several threads in the development of the field of sustainability science, drawing heavily on the engineering perspectives of the Integrated Research System for Sustainability Science Program at the University of Tokyo. It proposes steps for further development of the field and for the training of scholars and practitioners to contribute to it.

Chapter 4. Governance in Social-Environmental Systems

1. Acemoglu, D., J. A. Robinson, and D. Woren. 2012. *Why Nations Fail: The Origins of Power, Prosperity, and Poverty*. New York: Crown Business. 529 pages.

    This book analyzes the drivers of human development from a historical perspective. With the support of evidence from comparative case studies, the authors present a compelling argument about the importance of inclusive institutions in the governance of development efforts in society. Although the work focuses much more on economic development, it is directly relevant to our book's discussion of governance in social-environmental systems in that it offers ideas about why some societies are able to perform better than others in the pursuit of sustainability.

2. Gibson, C., K. Andersson, E. Ostrom, and S. Shivakumar. 2005. *The Samaritans' Dilemma: The Political Economy of Development Aid.* Oxford, UK: Oxford University Press. 264 pages.

Why is it so difficult to design and implement effective interventions for helping poor people in developing countries? How come so many foreign-aid projects fail to contribute to sustainable development? This book seeks to answer these questions through an institutional approach. In doing so it pays particular attention to the institutional arrangements within foreign-aid systems and argues that these arrangements often produce incentives inconsistent with the official goals of the interventions. The theoretical framework in the book is the main source of inspiration for the governance framework we present in chapter 4. This book will be of interest to readers interested in exploring further the analysis of how governance, institutional arrangements, and incentives influence human decisions and actions on the ground.

3. Grindle, M. S. 2004. "Good Enough Governance: Poverty Reduction and Reform in Developing Countries." *Governance* 17(4): 525–548.

Many scholars, practitioners, and policy makers who work on sustainable development call for "good governance" as one of the most important ingredients of the development process. In this article, Grindle challenges us all to be more precise and realistic in our definition of what "good governance" means. To be meaningful, she argues, it is necessary to define the *most* important attributes of the governance process in particular contexts. This context-sensitive approach to governance analysis will be particularly useful to readers interested in how governance affects sustainable development efforts in the developing world, where formal governmental institutions are often weak.

4. Poteete, A. R., M. A. Janssen, and E. Ostrom. 2010. *Working Together: Collective Action, the Commons, and Multiple Methods in Practice.* Princeton, NJ: Princeton University Press. 346 pages.

This book makes a compelling case for interdisciplinary scholarship to advance our current understanding of common-pool resources and how these might be governed more effectively. It provides an excellent overview of the many methodological challenges that face scholars working on issues related to the "commons" and how they can be addressed by scholars cooperating more effectively, not just with one another but also with the common-pool resource users they study.

5. Young, O. R. 2013. *On Environmental Governance: Sustainability, Efficiency, and Equity*. Boulder, CO: Paradigm Publishers. 196 pages.

   In this book one of the intellectual leaders in the field of environmental governance summarizes his life's work. It is written in a very accessible language, avoiding unnecessary jargon without oversimplifying the message. It is particularly helpful in clarifying the concept of multilevel governance its challenges, and discusses how it can help address some of our time's most pressing social-environmental problems.

Chapter 5. Linking Knowledge with Action

1. van Kerkhoff, L., and L. Lebel. 2006. "Linking Knowledge and Action for Sustainable Development." *Annual Review of Environment and Resources* 31: 445–477.

   This paper provides a masterful conceptual framework for thinking about linking knowledge with action for sustainable development and uses that framework to organize its review of the extensive literature on the subject. It is particularly strong on the relationships between knowledge and power, with telling examples of how even the best-intentioned researchers can inadvertently end up serving the interests of the powerful over those of the powerless.

2. National Research Council. Roundtable on Science and Technology for Sustainability. 2006. *Linking Knowledge with Action for Sustainable Development: The Role of Program Managers*. Washington, DC: National Academy Press.

   This document reports on a study conducted under the auspices of the National Academy of Sciences of the United States to determine how efforts to link knowledge with action have actually fared in research conducted and supported by the U.S. government. Government program managers with reputations for excellence in such linkage activities were brought together to identify common barriers to successful linkage activities and to share strategies for overcoming those barriers.

3. DeFries, R. 2014. *The Big Ratchet: How Humanity Thrives in the Face of Natural Crisis*. New York: Basic Books. 273 pages.

This original and accessible book tracks through humanity's time on Earth people's ability to manipulate nature. Its exposition ranges from the ability to control fire to expertise in breeding palatable plants, from the capacity to ship fertilizer across the Atlantic to skill in selectively tinkering with plant genomes. The author presents an even-handed treatment of how the abilities to reshape nature have both advanced and endangered human well-being.

4. Clark, W.C., P.A. Matson, and N. M. Dickson. 2008. "Knowledge Systems for Sustainable Development." Sackler Colloquium of the United States National Academy of Sciences. www.nasonline.org /SACKLER_sustainable_development.

This special feature reports on an integrated approach to understanding and promoting the mobilization of knowledge in pursuit of sustainability. Its empirical studies cover topics ranging from essential medicines to low-impact agriculture. It also provides a practical framework for guiding efforts to promote more effective research in support of sustainability.

Chapter 6. Next Steps: Contributing to a Sustainability Transition

The following are Web resources for keeping up with current developments in the practice of sustainable development.

1. United Nations. Sustainable Development Knowledge Platform. Access to international committees, positions of national and NGO groups, history of international summits, and the like: sustainable development.un.org/index.html.

2. World Business Council for Sustainable Development. The business community's central site for solutions, studies, and positions: www.wbcsd.org/about.aspx.

3. International Institute for Sustainable Development. The great ag-

gregator of NGO work around the world, as well as a substantial policy presence of its own: www.iisd.org.

4. SciDev.net. A truly global and slightly edgy site for bringing science and development together through news and analysis: www .scidev.net/global.

Notes

CHAPTER 1. PURSUING SUSTAINABILITY: AN INTRODUCTION

1. World Commission on Environment and Development (The Brundtland Commission). 1987. *Our Common Future*. New York: United Nations. www.un-documents.net/wced-ocf.htm.

2. UN Sustainable Development Knowledge Platform. Sustainable Development Goals. sustainabledevelopment.un.org/topics; Transforming our World: the 2030 Agenda for Sustainable Development. sustainabledevelopment.un.org/post 2015/transformingourworld.

CHAPTER 2. A FRAMEWORK FOR SUSTAINABILITY ANALYSIS: LINKING ULTIMATE GOALS WITH THEIR UNDERLYING DETERMINANTS

1. Dasgupta, P. 2004. *Human Well-Being and Natural Environment*. New York: Oxford University Press; United Nations International Human Dimensions Programme (UNU-IHDP) and United Nations Environment Programme (UNEP). 2012. *Inclusive Wealth Report 2012: Measuring Progress toward Sustainability*. Cambridge: Cambridge University Press.

2. International Energy Agency. 2013. Key World Energy Statistics 2013. www.iea.org/newsroomandevents/news/2013/october/key-world-energy-statistics-2013-now-available.html; Moomaw, W., T. Griffin, K. Kurczak, and J. Lomaz. 2012. *The Critical Role of Global Food Consumption Patterns in Achieving Sustainable Food Systems and Food for All*. A UNEP Discussion Paper. Paris: United Nations Environment Programme, Division of Technology, Industry and Economics; and United Nations Environment Programme (UNEP). 2008. "Trends in Global Water Use by Sector." In *Vital Water Graphics: An Overview of the State of the World's Fresh and Marine Waters*, 2nd ed. www.unep.org/dewa/vitalwater/article43.html.

3. See note 1, UNU-IHDP, UNEP.

4. See, for example, note 1, Dasgupta.

5. This box is drawn from United States Environmental Protection Agency (U.S. EPA) www.epa.gov/environmentaljustice/index.html; Natural Resources Defense

Council, "The Environmental Justice Movement," www.nrdc.org/ej/history/hej.asp; Cutter, S. L. 1995. "Race, Class and Environmental Justice." *Progress in Human Geography* 19: 111–122; and Iles, A. 2004. "Mapping Environmental Justice in Technology Flows: Computer Waste Impacts in Asia." *Global Environmental Politics* 4(4): 76–107. doi:10.1162/glep.2004.4.4.76.

6. McDonald, D. A., ed. 2002. *Environmental Justice in South Africa.* Athens: Ohio University Press.

7. Official Journal of the European Union. 2000. "Charter of Fundamental Rights of the European Union." Article 37. ec.europa.eu/justice/fundamental-rights /charter.

8. An example of a politically driven, multidimensional survey approach to well-being is provided by a recent Organization for Economic Cooperation and Development (OECD) project called the "Better Life Index." OECD posts the resulting data on a user-friendly Web site, aiming to "engage people in the debate on well-being" and to highlight what needs to be done to improve it. Users can apply their own weights, examine trade-offs and equity issues, and—presumably—improve society's capacity to reflect on what constitutes "the good life." OECD has subsequently extended its analysis in both time and space to document historical trends in constituents of well-being for 80% of the world since the early nineteenth century. www.oecd.org/statistics/datalab/bli.htm.

9. McGregor, J. Allister, L. Camfield, and A. Woodcock. 2009. "Needs, Wants and Goals: Wellbeing, Quality of Life and Public Policy." *Applied Research in Quality of Life* 4(2): 143.

10. World Health Organization (WHO). 2005. *Ecosystems and Human Well-Being: Health Synthesis*, vol. 5. www.who.int/entity/globalchange/ecosystems/ecosys.pdf.

11. Sen, A. 2013. "The Ends and Means of Sustainability." *Journal of Human Development and Capabilities* 14(1): 7.

12. Kearney, J. 2010. "Food Consumption Trends and Drivers." *Philosophical Transactions of the Royal Society B: Biological Sciences* 365(1554): 2793–807.

13. Food and Agriculture Organization of the United Nations (FAO), International Fund for Agricultural Development (IFAD) and World Food Programme (WFP). 2015. *The State of Food Insecurity in the World: Meeting the 2015 International Hunger Targets; Taking Stock of Uneven Progress.* Rome: FAO. www.fao.org/3/a -i4646e/index.html.

14. WHO. 2015. "World Health Statistics 2015." www.who.int/publications/en.

15. United Nations Inter-agency Group for Child Mortality Estimation (UN IGME). 2015. *Levels and Trends in Child Mortality*. Report 2015. www.childmortal ity.org.

16. Lozano, R., M. Naghavi, K. Foreman, S. Lim, K. Shibuya, V. Aboyans, J. Abraham, et. al. 2012. "Global and Regional Mortality from 235 Causes of Death for 20 Age Groups in 1990 and 2010: A Systematic Analysis for the Global Burden of Disease Study 2010." *Lancet* 380(9859): 2095–128. doi:10.1016/S0140–6736(12) 61728–0.

17. See the preceding note.

18. Reddy, K. S., and S. Yusuf. 1998. "Emerging Epidemic of Cardiovascular Disease in Developing Countries." *Circulation* 97(6): 596–601.

19. Lawes, C.M.M., S. V. Hoorn, and A. Rodgers. 2008. "Global Burden of Blood-Pressure-Related Disease, 2001." *Lancet* 371(9623): 1513–18. doi:10.1016/S0140–6736(08)60655–8.

20. Sicree, R., and J. Shaw. 2007. "Type 2 Diabetes: An Epidemic or Not, and Why It Is Happening." *Diabetes & Metabolic Syndrome: Clinical Research & Reviews* 1(2): 75–81. doi:10.1016/j.dsx.2006.11.012.

21. See note 18.

22. World Health Organization (WHO). 2013. *The Top 10 Causes of Death.* who.int/mediacentre/factsheets/fs310/en/index1.html.

23. Ezzati, M., A. D. Lopez, A. Rodgers, S. V. Hoorn, and C. J. L. Murray. 2002. "Selected Major Risk Factors and Global and Regional Burden of Disease." *Lancet* 360(9343): 1347–60. doi:10.1016/S0140–6736(02)11403–6.

24. See note 15, UN IGME, 7.

25. The data and figure on education are taken from UNESCO Institute for Statistics. 2014. Adult and Youth Literacy Fact Sheet. UIS Fact Sheet No. 29. www.uis.unesco.org/literacy/Documents/fs-29-2014-literacy-en.pdf.

26. See note 1, UNU-IHDP, UNEP, 234.

27. Credit Suisse. 2014. *Credit Suisse Global Wealth Databook 2014.* Zurich: Credit Suisse Research Institute. publications.credit-suisse.com/tasks/render/file/?fileID =5521F296-D460-2B88-081889DB12817E02.

28. United Nations Development Programme (UNDP). 2010. *Regional Human Development Report for Latin America and the Caribbean 2010.* hdr.undp.org/en/reports.

29. UNDP. 2012. "Empowering Women Is Key to Building a Future We Want, Nobel Laureate Says." *UNDP News Centre.* www.undp.org/content/undp/en/home/presscenter/articles/2012/09/27/empowering-women-is-key-to-building-a-future -we-want-nobel-laureate-says.html.

30. Cutter, S. L., L. Barnes, M. Berry, C. Burton, E. Evans, E. Tate, and J. Webb. 2008. "A Place-Based Model for Understanding Community Resilience to Natural Disasters." *Global Environmental Change* 18(4): 598–606.

31. UNDP. *Human Development Report 2013.* "The Rise of the South: Human Progress in a Diverse World." hdr.undp.org/en/2013-report, 38.

32. Millennium Ecosystem Assessment. 2005. *Ecosystems and Human Well-Being, Current State and Trends*, vol. 1, edited by Hassan, R., R. Scholes, and N. Ash. Washington, DC: Island Press. www.unep.org/maweb/en/Condition.aspx#download.

33. Intergovernmental Platform on Biodiversity and Ecosystem Services (IPBES). www.ipbes.net.

34. See note 1, UNU-IHDP and UNEP.

35. Biermann, F. 2014. "The Anthropocene: A Governance Perspective." *The Anthropocene Review* 1(1): 57–61. doi:10.1177/2053019613516289.

36. OECD. 2010. "Framing Eco-Innovation: The Concept and Evolution of

Sustainable Manufacturing." In *Eco-Innovation in Industry: Enabling Green Growth.* dx.doi.org/10.1787/9789264077225-4-en.

37. This box draws on McDonough, W., and M. Braungart. 2010. *Cradle to Cradle: Remaking the Way We Make Things.* New York: Macmillan; McDonough, W., and M. Braungart. 2003. "Towards a Sustaining Architecture for the 21st Century: The Promise of Cradle-to-Cradle Design." *Industry and Environment* 26(2): 13–16; and Rossi, M., S. Charon, G. Wing, and J. Ewell. 2006. "Design for the Next Generation: Incorporating Cradle-to-Cradle Design into Herman Miller Products." *Journal of Industrial Ecology* 10(4): 193–210.

38. McDonough Braungart Design Chemistry. www.mbdc.com.

39. See note 37, McDonough and Braungart 2003, 14.

40. "Human capital" draws for its conceptual approach on Cohen, J. E. 2010. "Beyond Population: Everyone Counts in Development." *Center for Global Development Working Paper* 220, Washington, DC: Center for Global Development. www.cgdev.org/content/publications/detail/1424318. Data are from Population Reference Bureau. 2014. *World Population Data Sheet 2014.* Washington, DC: Population Reference Bureau.

41. Hancock, T. 2001. "People, Partnerships and Human Progress: Building Community Capital." *Health Promotion International* 16(3): 276. doi:10.1093/heapro/16.3.275.

42. Pelling, M., and C. High. 2005. "Understanding Adaptation: What Can Social Capital Offer Assessments of Adaptive Capacity?" *Global Environmental Change* 15(4): 308–19. doi:10.1016/j.gloenvcha.2005.02.001.

CHAPTER 3. DYNAMICS OF SOCIAL-ENVIRONMENTAL SYSTEMS

1. The complex systems we address here have been referred to in the literature as socio-ecological systems, human-environment systems, coupled human-natural systems, and social-environmental systems. We use the latter term in this book because we consider it the most inclusive of the alternatives. In particular, "environment" is a term that includes ecological systems as well as climate, atmosphere, hydrologic, and rock and mineral resource systems. And "social" emphasizes our interest in the development of society as a whole, not just the individual humans that constitute society's members.

2. Pepys, Samuel. 1926. *Everybody's Pepys: The Diary of Samuel Pepys 1600–1669,* edited by O. F. Morshead, 395. New York: Harcourt, Brace.

3. Intergovernmental Panel on Climate Change. www.ipcc.ch.

4. The quote has been attributed to Garrett Hardin of "*The Tragedy of the Commons*" fame (see chap. 4). Less succinct, but more lyrical is John Muir's "When we try to pick out anything by itself, we find it hitched to everything else in the Universe," from his book: *My First Summer in the Sierra.* 1911. Boston: Houghton Mifflin.

5. For more on tipping points, see Lenton, T. M. 2013. "Environmental Tipping Points." *Annual Review of Environment and Resources* 38(1): 1–29. doi:10.1146

/annurev-environ-102511–084654; Biggs, R., T. Blenckner, C. Folke, L. Gordon, A. Norström, M. Nyström, G. Peterson. 2012. "Regime Shifts." In *Encyclopedia of Theoretical Ecology,* A. Hastings and I. Gross, eds., 609–16. University of California Press; Carpenter, S. R. 2003. *Regime Shifts in Lake Ecosystems: Pattern and Variation.* Oldendorf, Germany: International Ecology Institute; Lenton, T. M., H. Held, E. Kriegler, J. W. Hall, W. Lucht, S. Rahmstorf, and H. J. Schellnhuber. 2008. "Tipping Elements in the Earth's Climate System." *Proceedings of the National Academy of Sciences of the United States of America* 105(6): 1786–93; Carter, M. R., and C. B. Barrett. 2006. "The Economics of Poverty Traps and Persistent Poverty: An Asset-Based Approach." *Journal of Development Studies* 42(2): 178–99; and Barrett, C. B., A. J. Travis, and P. Dasgupta. 2001. "On Biodiversity Conservation and Poverty Traps." *Proceedings of the National Academy of Sciences of the United States of America* 108(34): 13907–12.

   6. See, for example: Lenton, T. M., "Environmental Tipping Points," in the preceding note.

   7. See note 5, Carpenter

   8. See note 5, Lenton et al., "Tipping Elements."

   9. See note 5, Carter and Barrett.

10. Biggs, R., M. Schlüter, D. Biggs, E. L. Bohensky, S. Burnsilver, G. Cundill, V. Dakos, T. Daw, L. Evans, K. Kotschy, A. Leitch, C. Meek, A. Quinlan, C. Raudsepp-Hearne, M. Robards, M. L. Schoon, L. Schultz, and P. C. West. 2012. "Toward Principles for Enhancing the Resilience of Ecosystem Services." *Annual Review of Environment and Resources* 37(1): 421–48.

11. See the preceding note.

12. For more information about ecosystem services models in action, see the natural capital project, ww.naturalcapitalproject.org, and ProEcoServ, www.proeco serv.org. For more detailed reading, see Crossman, N. D., B. Burkhard, S. Nedkov, L. Willemen, K. Petz, I. Palomo, E. G. Drakou, B. Martín-Lopez, T. McPhearson, K. Boyanova, R. Alkemade, B. Egoh, M. B. Dunbar, and J. Maes. 2013. "A Blueprint for Mapping and Modelling Ecosystem Services." *Ecosystem Services*, Special Issue on Mapping and Modelling Ecosystem Services, 4(June): 4–14. See content on econometrics at MIT OpenCourseWare: ocw.mit.edu/index.htm. Read more about LCA models in Hendrickson, C. T., L. B. Lave, and H. S. Matthews. 2010. *Environmental Life Cycle Assessment of Goods and Services: An Input-Output Approach.* Washington, DC: RFF Press. For examples of current research utilizing LCA models, see Stanford University's Environmental Assessment and Optimization Group: pangea.stanford .edu/researchgroups/eao/research/life-cycle-assessment.

13. This case draws on Goldstein, J. H., G. Caldarone, C. Colvin, T. K. Duarte, D. Ennaanay, K. Fronda, N. Hannahs, E. McKenzie, G. Mendoza, K. Smith, S. Wolny, U. Woodside, and G. C. Daily. 2010. "TEEB case: Integrating Ecosystem Services into Land-Use Planning in Hawai'i, USA." Available at www.TEEBweb.org; and materials at www.naturalcapitalproject.org.

14. Klöpffer, W. 1997. "Life Cycle Assessment." *Environmental Science and Pollution Research* 4(4): 223–28.

15. Guinee, J. B., R. Heijungs, G. Huppes, A. Zamagni, P. Masoni, R. Buonamici, T. Ekvall, and T. Rydberg. 2010. "Life Cycle Assessment: Past, Present, and Future." *Environmental Science & Technology* 45(1): 90–96.

16. Parris, T. M., and R. W. Kates. 2003. "Characterizing and Measuring Sustainable Development." *Annual Review of Environment and Resources* 28: 559–86.

17. Muller, N. Z., R. Mendelsohn, and W. Nordhaus. 2011. "Environmental Accounting for Pollution in the United States Economy." *American Economic Review* 101: 1649–75.

18. The Inclusive Wealth Project: inclusivewealthindex.org.

19. UNU-IHDP, UNEP. 2014. *Inclusive Wealth Report 2014: Measuring Progress toward Sustainability.* Cambridge: Cambridge University Press.

CHAPTER 4. GOVERNANCE IN SOCIAL-ENVIRONMENTAL SYSTEMS

1. For our definition and conceptual discussion of governance, we draw on Grindle, M. S. 2004. "Good Enough Governance: Poverty Reduction and Reform in Developing Countries." *Governance* 17(4): 525–48; Andersson, K., G. Gordillo, and F. van Laerhoven. 2009. *Local Governments and Rural Development: Comparing Lessons from Brazil, Chile, Mexico, and Peru.* Tucson: University of Arizona Press.

2. For details on the story of how this agreement came to fruition, please see Hoen, E., J. Berger, A. Calmy, S. Moon. 2011. "Driving a Decade of Change: HIV/AIDS, Patents, and Access to Medicines." *Journal of the International AIDS Society* 14:15; and in Hein, W., and S. Moon. 2013. *Informal Norms in Global Governance: Human Rights, Intellectual Property and Access to Medicines.* Aldershot, UK: Ashgate.

3. Hardin, G. 1968. "The Tragedy of the Commons." *Science* 162(3859): 1243–48.

4. For a more in-depth discussion of how incentives affect sustainable-development efforts, see Gibson, C., K. Andersson, E. Ostrom, and S. Shivakumar. 2005. *The Samaritans' Dilemma: The Political Economy of Development Aid.* Oxford, UK: Oxford University Press.

5. Liverman, D. M., and S. Vilas. 2006. "Neoliberalism and the Environment in Latin America." *Annual Review of Environment and Resources* 31: 327–63.

6. World Commission on Dams. 2000. *Dams and Development: A New Framework for Decision-Making: The Report of the World Commission on Dams.* London: Earthscan.

7. Jäger, J., N. Dickson, A. Fenech, P. Haas, E. Parson, V. Sokolov, F. Toth, J. van der Sluijs, and C. Waterton. 2001. "Monitoring in the Management of Global Environmental Risks." Chapter 16 in Social Learning Group. 2001. *Learning to Manage Global Environmental Risks, Volume 2: A Functional Analysis of Social Responses to Climate Change, Ozone Depletion, and Acid Rain,* edited by W. Clark, J. Jäger, J. van Eijndhoven, and N. Dickson. Cambridge, MA: MIT Press.

8. Frey, B. S., and F. Oberholzer-Gee. 1997. "The Cost of Price Incentives: An Empirical Analysis of Motivation Crowding-Out." *American Economic Review* 87(4): 746–55.

9. Gneezy, U., and A. Rustichini. 2000. "Pay Enough or Don't Pay At All." *Quarterly Journal of Economics* 115(3): 791–810.

10. Cardenas, J. C., J. Stranlund, and C. Willis. 2000. "Local Environmental Control and Institutional Crowding-Out." *World Development* 28(10): 1719–33.

11. Ostrom, E. 1990. *Governing the Commons: The Evolution of Institutions for Collective Action.* Cambridge: Cambridge University Press.

12. For a review of 91 such empirical studies, see Cox, M., G. Arnold, and S. Villamayor Tomás. 2010. "A Review of Design Principles for Community-Based Natural Resource Management." *Ecology and Society* 15(4): 38.

13. Somanathan, E., R. Prabhakar, and B. Mehta. 2009. "Decentralization for Cost-Effective Conservation." *Proceedings of the National Academy of Sciences of the United States of America* 106(11): 4143–47.

14. For example, see Young, O. R. 2013. *On Environmental Governance: Sustainability, Efficiency, and Equity.* Boulder, CO: Paradigm; Andersson, K., and E. Ostrom. 2008. "Analyzing Decentralized Natural Resource Governance from a Polycentric Perspective." *Policy Sciences* 41(1): 1–23.

15. See Ostrom, E. 2010. "Polycentric Systems for Coping with Collective Action and Global Environmental Change." *Global Environmental Change* 20(4): 550–57.

CHAPTER 5. LINKING KNOWLEDGE WITH ACTION

1. This chapter draws heavily on the Senator George J. Mitchell Lectures on Sustainability presented at University of Maine, Orono, ME by Matson, P.A. 2012. "A Call to Arms for a Transition to Sustainability," vimeo.com/51780215; and by Clark, W.C. 2014. "Mobilizing Knowledge to Shape a Sustainable Future," vimeo.com/112854984.

2. Cash, D. W., W. C. Clark, F. Alcock, N. M. Dickson, N. Eckley, D. H. Guston, J. Jäger, and R. B. Mitchell. 2003. "Knowledge Systems for Sustainable Development." *Proceedings of the National Academy of Sciences of the United States of America* 100(14): 8086–91. doi:10.1073/pnas.1231332100.

3. This story was told by Julia Novy-Hildesley, a lecturer at Stanford University. She served for a decade as executive director of the Lemelson Foundation, where she spearheaded the foundation's program in user-driven technology design and dissemination in Asia, Africa, and Latin America.

4. Ahuja, A., M. Kremer, and A. P. Zwane. 2010. "Providing Safe Water: Evidence from Randomized Evaluations." *Annual Review of Resource Economics* 2: 237–56.

5. National Oceanic and Atmospheric Administration (NOAA). 2013. About the Regional Integrated Sciences and Assessments Program. cpo.noaa.gov/Climate Programs/ClimateandSocietalInteractions/RISAProgram/AboutRISA.aspx; Buizer,

J., K. Jacobs, and D. Cash. 2010. "Making Short-Term Climate Forecasts Useful: Linking Science and Action." *Proceedings of the National Academy of Sciences of the United States of America.* doi:10.1073/pnas.0900518107.

6. Hellström, T., and M. Jacob. 2003. "Boundary Organizations in Science: From Discourse to Construction." *Science and Public Policy* 30(4): 235–38.

7. Palm, C. A., S. A. Vosti, P. A. Sanchez, and P. J. Ericksen. 2005. *Slash-and-Burn Agriculture: The Search for Alternatives.* New York: Columbia University Press.

8. Clark, W. C., T. P. Tomich, M. van Noordwijk, D. Guston, D. Catacutan, N. M. Dickson, and E. McNie. 2011. "Boundary work for sustainable development: Natural resource management at the Consultative Group on International Agricultural Research (CGIAR)." *Proceedings of the National Academy of Sciences of the United States of America.* doi:10.1073/pnas.0900231108.

9. Jones, N., H. Jones, and C. Walsh. 2008. "Political Science? Strengthening Science-Policy Dialogue in Developing Countries." Working Paper 294, Overseas Development Institute, London; Mollinga, P. P. 2010. "Boundary Work and the Complexity of Natural Resources Management." *Crop Science* 50: S-1-S-9.

CHAPTER 6. NEXT STEPS: CONTRIBUTING TO A SUSTAINABILITY TRANSITION

1. National Research Council (NRC). 1999. *Our Common Journey: A Transition to Sustainability.* NRC Board on Sustainable Development. Washington, DC: National Academy Press.

2. The story about Maria Foronda is based on the biographic information about past winners of the Goldman Environmental Prize—an award she won in 2003. goldmanprize.org.

3. Nijhus, M. 2003. "A Peruvian Activist Takes on the Fishmeal Industry." April 18, 2003. Grist.org.

4. Young, S. 2011. "Ray Anderson: Climbing Mount Sustainability; A Case Study about Ethical Leadership (teaching materials)."

5. Rosenberg, B. 2009. "Interface Carpet and Fabric Company's Sustainability Efforts: What the Company Does, the Crucial Role of Employees, and the Limits of This Approach." *Journal of Public Health Policy* 30(4): 427–38.

6. See the preceding note.

7. Minter, S. 2013. "A Net Gain for Sustainable Manufacturing." *IndustryWeek. com*, August 2013.

8. See note 4.

9. Lacy, P., and J. Rutqvist. 2015. *Waste to Wealth: The Circular Economy Advantage.* London: Palgrave Macmillan.

10. Sen, Amartya. 2013. "The Ends and Means of Sustainability." *Journal of Human Development and Capabilities* 14(1): 6–20.

APPENDIX A. CASE STUDIES IN SUSTAINABILITY

1. This case study is taken from the longer and more fully documented account in Clark, W. C. 2015. "London: A Multi-Century Struggle for Sustainable Development in an Urban Environment." HKS Faculty Research Working Paper Series No. RWP15-047. Harvard Kennedy School of Government. Cambridge, MA. research .hks.harvard.edu/publications/workingpapers/citation.aspx?PubId = 9812&type = FN&PersonId = 124.

2. IESE Center for Globalization and Strategy. 2014. *Cities in Motion Index 2014*. University of Navarra, Spain: IESE Business School; PricewaterhouseCoopers (PwC). 2014. Cities of Opportunity (No. 6). Delaware, USA: PricewaterhouseCoopers; Economist Intelligence Unit. 2012. *The Green City Index*. Munich, Germany: Siemens.

3. Greater London Authority. 2011. "Context and Strategy" (chap. 1, para. 1.52). In *The London Plan: Spatial Development Strategy for Greater London July 2011*. London: Greater London Authority.

4. Brimblecombe, P. 1987. *The Big Smoke: A History of Air Pollution in London since Medieval Times*. London, New York: Methuen, 68. The author also notes that London had none of the inversion-prone basin attributes that would eventually so plague the air pollution control efforts of other cities, such as Los Angeles.

5. Glacken, C. J. 1967. *Traces on the Rhodian Shore: Nature and Culture in Western Thought from Ancient Times to the End of the Eighteenth Century*. Berkeley, CA: University of California Press, 336.

6. This section draws extensively on the following accounts: For water, Halliday, S. 1999. *The Great Stink of London: Sir Joseph Bazalgette and the Cleansing of the Victorian Capital*. Thrupp, Stroud, Gloucestershire: Sutton; and Green, C. H. 2010. *Case Study Brief: Sustainable Urban Water Management in London. (Input to deliverable 6.1.5–6 Comparative Analysis of Enabling Factors for Sustainable Urban Water Management)* Switch Project. For forest, coal, and air quality connections, Tebrake, W. H. 1975. "Air-Pollution and Fuel Crises in Preindustrial London, 1250–1650." *Technology and Culture* 16(3): 337–59.

7. See the preceding note, Halliday, p. 44, quoting H. T. Riley, ed., *Memorials of London Life*, 295. Metropolitan Archives.

8. See note 6, Tebrake, citing Duby, G. 1971. "Medieval Agriculture 900–1500." In *The Fontana Economic History of Europe*, edited by C. M. Cipolla, 199. London: Collins.

9. See note 6, Tebrake, citing *Calendar of Close Rolls, Edward I (1302–7)*, 537.

10. Information on the plague in London is from Museum of London. 2011. *London Plagues 1348–1665*. www.museumoflondon.org.uk/explore-online/pocket -histories/london-plagues-13481665/black-death-13481350.

11. Galloway, J. A., and J. S. Potts. 2007. "Marine Flooding in the Thames Estuary and Tidal River c.1250–1450: Impact and Response." *Area* 39(3): 370–79.

12. See note 9.

13. 23 Henry VIII Cap. V; This and related measures were apparently effective. Jenner has shown that refuse disposal in sixteenth- and seventeenth-century London was generally well organized; officials required that streets be swept twice per day, at least in central London (Jenner, M.S.R. 1991. "Early Modern English Conceptions of 'Cleanliness' and 'Dirt' as Reflected in the Environmental Regulation of London, c. 1530–c. 1700," [Ph.D. diss., Oxford University], especially chap. 2.)

14. See note 4, Brimblecombe, citing the work of seventeenth-century scientists Margaret Cavendish, Kenelm Digby, and John Evelyn.

15. See note 6, Halliday, quoting *Analytical Index to the Rememgrencia, 1579–1664* (Corporation of London, 1878), 482. Guildhall Library.

16. Malnutrition was endemic in London, especially for the poor. (Hard data are not available until later periods; however, poor children born in London in the mid-eighteenth century would grow up to be a full 10 inches shorter than their contemporaries in rural England or than children born in London two centuries later. That said, there is little convincing evidence that changes in nutritional status were associated with major disease outbreaks. See Landers, J. 1993. *Death and the Metropolis: Studies in the Demographic History of London, 1670–1830*, 66–68. Cambridge, New York: Cambridge University Press.

17. Twigg, G. 1993. "Plague in London: Spatial and Temporal Aspects of Mortality." In *Epidemic Disease in London* (*Center for Metropolitan History Working Paper Series No. 1*), edited by J.A.I. Champion, 1–17. London: Center for Metropolitan History. events.sas.ac.uk/support-research/publications/923.

18. See note 10.

19. The account of the Great Fire and its consequences is from Museum of London. 2005. *London's Burning: The Great Fire of London 1666.* museumoflondon.org.uk/explore-online/pocket-histories/what-happened-great-fire-london.

20. See the preceding note.

21. Finlay, R. 1981. *Population and Metropolis: The Demography of London, 1580–1650.* Cambridge, New York: Cambridge University Press.

22. George, M. D. 1965. *London Life in the Eighteenth Century*, 35–36. New York: Harper & Row.

23. See note 16, Landers, 47.

24. A thorough case for the role of breastfeeding in the high mortality rates of eighteenth-century London is made in Landers, *Death and the Metropolis* (see note 16), 47. Further insights into the data behind this case are given in King, S. 1997. "Dying with Style: Infant Death and Its Context in a Rural Industrial Township 1650–1830." *Social History of Medicine* 10(1): 3–24; and Knodel, J., and H. Kinter. 1977. "Impact of Breastfeeding Patterns on Biometric Analysis of Infant-Mortality." *Demography* 14(4): 391–409. The assertion that breastfeeding can increase infant survival chances five- to tenfold is from a study by Huck (1997) of the generally more sanitary and less disease-infested London of 1908–1918, and thus almost certainly represents a lower limit (Huck, P. 1997. "Shifts in the Seasonality of Infant

Deaths in Nine English Towns during the 19th Century: A Case for Reduced Breast Feeding?" *Explorations in Economic History* 34(3): 368–86). An accessible summary of the benefits of breastfeeding is provided in Jackson, K. M., and A. M. Nazar. 2006. "Breastfeeding, the Immune Response, and Long-Term Health." *Journal of the American Osteopathic Association* 106(4): 203–7.

25. The progress is most evident in the wealthy, who adopted variolation early. For England as a whole, until the middle of the eighteenth century life expectancy of ducal families was indistinguishable from that of commoners, averaging 30–35 years. After that, being rich began to matter: by 1866–71, a child born into a ducal family could expect to live sixty years, half again as long as the average Englander. See Harris, B. 2004. "Public Health, Nutrition, and the Decline of Mortality: The McKeown Thesis Revisited." *Social History of Medicine* 17(3): 379–407.

26. The term *vaccination* was introduced by scientist Edward Jenner to describe his demonstration in 1796 that by exposing a person to cowpox—a relatively mild disease commonly experienced by milkmaids—he could protect them from subsequent infections by smallpox. Jenner's paper reporting the evidence for his new approach was rejected by the Royal Society. He nevertheless persisted, privately published a report on his findings, and distributed the vaccine freely to all who were willing to try it for themselves or to treat others. Despite much controversy—part of it arising from campaigns in support of the then-well-established but less effective variolation procedure—he and his followers achieved results. By the early nineteenth century widespread acceptance of vaccination brought about rapid declines in smallpox deaths.

27. Principal sources for this smallpox story are Razzell, P. E. 2003. *The Conquest of Smallpox: The Impact of Inoculation on Smallpox Mortality in Eighteenth Century Britain*, 2nd ed. London: Caliban Books; and Riedel, S. 2005. "Edward Jenner and the History of Smallpox and Vaccination." *Proceedings Baylor University Medical Center* 18(1): 21–25. Useful additional perspectives drawing on newer data are provided in Davenport, R., L. Schwarz, and J. Boulton. 2011. "The Decline of Adult Smallpox in Eighteenth-Century London." *Economic History Review* 64(4) 1289–1314; and Razzell, P. E. 2011. "The Decline of Adult Smallpox in Eighteenth-Century London: A Commentary." *Economic History Review* 64(4): 1315–35.

28. See note 16, Landers.

29. See note 6. This section draws heavily on Green, *Case Study Brief*; and Halliday, *The Great Stink of London*.

30. See note 6, Halliday, 52.

31. Simon, J. 1856. *Report of the Last Two Cholera-Epidemics of London, as Affected By the Consumption of Impure Water*, 12. London, England: General Board of Health.

32. See the preceding note, p. 10.

33. Daunton, M. 2004. *London's "Great Stink" and Victorian Urban Planning.* www.bbc.co.uk/history/trail/victorian_britain/social_conditions/victorian_urban_planning_01.shtml. (This page has been archived and is no longer being updated.)

34. See note 6, Halliday, 107.

35. See note 6. The innovation arguments advanced here are drawn from Green, *Case Study Brief*.

36. See note 4, Brimblecombe.

37. It is worth noting, in light of the present debates around indoor cookstoves in the developing world, that according to Brimblecombe (see note 4, p. 55), one observer in the early seventeenth century wrote "that the number of chimneys had increased greatly since his youth (mid-sixteenth century). In those times, he wrote, smoke indoors had been regarded as hardening the timbers of the house and as a disinfectant to ward off disease."

38. BBC. 2002. *Days of Toxic Darkness: Interview with Barbara Fewster.* news.bbc .co.uk/2/hi/uk_news/2542315.stm.

39. BBC. 1952. *London Fog Clears after Days of Chaos.* news.bbc.co.uk/onthis day/hi/dates/stories/december/9/newsid_4506000/4506390.stm.

40. Bell, M. L., and D. L. Davis. 2001. "Reassessment of the Lethal London Fog of 1952: Novel Indicators of Acute and Chronic Consequences of Acute Exposure to Air Pollution." *Environmental Health Perspectives Supplements* 109: 389–94.

41. Social Learning Group. 2001. *Learning to Manage Global Environmental Risks.* Vol. 1: *A Comparative History of Social Responses to Climate Change, Ozone Depletion, and Acid Rain.* Cambridge, MA: MIT Press.

42. Draws on narrative in Benjamin, P. "Farming in the Himalayas; Living in a Perilous Environment." In *Institutions, Incentives, and Irrigation in Nepal*, edited by P. Benjamin, W. F. Lam, E. Ostrom, and G. P. Shivakoti. U.S. Agency for International Development, Global Bureau Democracy Center.

43. Lam, Wai Fung. 1996. "Improving the Performance of Small-Scale Irrigation Systems: The Effects of Technological Investments and Governance Structure on Irrigation Performance in Nepal." *World Development* 24(8): 1301–15.

44. See note 42.

45. This section is drawn from Lam; see note 43.

46. Zebu oxen are a major household asset, and not all households are wealthy enough to own a pair (you need two to plow the fields). The owner will rent out a pair to other community members based on fixed daily rate.

47. Child under-five mortality rate in Nepal for 1970: 269 out of 1,000 (UNICEF. 2014. *Every Child Counts: Revealing Disparities, Advancing Children's Rights.* New York: United Nations. www.unicef.org/sowc2014/numbers.

48. See the preceding note.

49. This section is drawn from Lam, W. F., and E. Ostrom. 2010. "Analyzing the Dynamic Complexity of Development Interventions: Lessons from an Irrigation Experiment in Nepal." *Policy Sciences* 43(1): 1–25. doi:10.1007/s11077-009-9082-6; see also Pradhan, P. 1989. "Patterns of Irrigation Organization in Nepal: A Comparative Study of 21 Farmer-Managed Irrigation Systems." Colombo, Sri Lanka: International Irrigation Management Institute.

50. Yoder, R. 1994. "Locally Managed Irrigation Systems: Essential Tasks and

Implications for Assistance, Management Transfer, and Turnover Programs." Colombo, Sri Lanka: International Irrigation Management Institute.

51. Joshi, N. N., E. Ostrom, G. P. Shivakoti, and W. F. Lam. 2000. "Institutional Opportunities and Constraints in the Performance of Farmer-Managed Irrigation Systems in Nepal." *Asia-Pacific Journal of Rural Development* 10(2): 67–92.

52. As told by Pamela Matson, on behalf of the Yaqui research team. A complete discussion of the Yaqui Valley project can be found in Matson, P. A., ed. 2012. *Seeds of Sustainability: Lessons from the Birthplace of the Green Revolution in Agriculture.* Washington, DC: Island Press.

53. Matson, P. A., R. L. Naylor, and I. Ortiz-Monasterio. 1998. "Integration of Environmental, Agronomic, and Economic Aspects of Fertilizer Management." *Science* 280: 112–15.

54. Freidberg, S. E. 2009. *Fresh: A Perishable History.* Cambridge, MA: Harvard University Press.

55. Andersen, S. O., and K. M. Sarma. 2012. *Protecting the Ozone Layer: The United Nations History.* London: Earthscan.

56. Molina, M., and F. S. Rowland. 1974. "Stratospheric Sink for Chlorofluoromethanes: Chlorine Atom Catalyzed Destruction of Ozone." *Nature* 249: 810–12.

57. See note 55.

58. Tolba, M. K., and I. Rummel-Bulska. 1998. *Global Environmental Diplomacy: Negotiating Environmental Agreements for the World, 1973–1992.* Cambridge: MIT Press.

59. DeSombre, E. R. 2000. "The Experience of the Montreal Protocol: Particularly Remarkable, and Remarkably Particular." *UCLA Journal of Environmental Law and Policy* 19: 49.

60. Benedict, R. E. 1998. *Ozone Diplomacy: New Directions in Safeguarding the Planet,* 105. Cambridge: Harvard University Press.

61. See the preceding note.

62. World Meteorological Organization. 1985. *Atmospheric Ozone, 1985: Assessment of Our Understanding of the Processes Controlling Its Present Distribution and Change,* Geneva archive.org/details/nasa_techdoc_19860023425.

63. UNEP. 2011. "United Nations Environment Programme: Montreal Protocol." ozone.unep.org/new_site/en/montreal_protocol.php.

Index